The Happiness Riddle and the Quest for a Good Life

Mark Cieslik

The Happiness Riddle and the Quest for a Good Life

Mark Cieslik
Northumbria University,
Newcastle upon Tyne, United Kingdom

ISBN 978-0-230-28303-9 (hardcover) ISBN 978-1-137-31882-4 (eBook)
ISBN 978-1-349-59224-1 (softcover)
DOI 10.1057/978-1-137-31882-4

Library of Congress Control Number: 2016958025

Printed on acid-free paper

This Palgrave Macmillan imprint is published by Springer Nature
The registered company is Macmillan Publishers Ltd. London
The registered company address is: The Campus, 4 Crinan Street, London, N1 9XW, United Kingdom

Dedicated to two young people,
June Jones and Kazimierz Cieslik,
South Wales, 1946,
'where it all began'

Acknowledgements

This book has been a long time in the making so there are many people who have helped me and deserve some thanks. Colleagues at Northumbria University have offered support and constructive criticism since I first started writing this book. Likewise, many people from the BSA Happiness Study Group have contributed ideas and inspired me to keep working on this project. In particular, Laura Hyman and Alexandra Jugureanu have supported me over the years in my efforts to finish this text. I also thank the editorial staff at Palgrave for their patience and also the 19 interviewees who made this book possible. I also thank my various family members for tolerating the years and years of conversations about happiness and wellbeing. My friends too have also endured countless conversations about happiness that have helped make this a much better book. My gratitude goes to Steve Miles, Don Simpson, Craig Wood, Mike, Jason, Phil, Sharon, Gerald, Jayne and Rob. Finally, I thank Clare and Theo who have provided all manner of insight, support and love, without which this book would never have been finished.

Contents

1

Introduction: The Happiness Riddle

All the world's a stage,
And all the men and women merely players;
They have their exits and their entrances;
And one man in his time plays many parts,
His acts being seven ages. At first the infant,
Mewling and puking in the nurse's arms.
And then the whining school-boy, with his satchel
And shining morning face, creeping like snail
Unwillingly to school. And then the lover,
Sighing like furnace, with a woeful ballad
Made to his mistress' eyebrow. Then a soldier,
Full of strange oaths and bearded like the pard,
Jealous in honour, sudden and quick in quarrel,
Seeking the bubble reputation
Even in the cannon's mouth. And then the justice,
In fair round belly with good capon lined,
With eyes severe and beard of formal cut,
Full of wise saws and modern instances;
And so he plays his part. The sixth age shifts

M. Cieslik, *The Happiness Riddle and the Quest for a Good Life*,
DOI 10.1057/978-1-137-31882-4_1

Into the lean and slipper'd pantaloon,
With spectacles on nose and pouch on side,
His youthful hose, well saved, a world too wide
For his shrunk shank; and his big manly voice,
Turning again toward childish treble, pipes
And whistles in his sound. Last scene of all,
That ends this strange eventful history,
Is second childishness and mere oblivion,
Sans teeth, sans eyes, sans taste, sans everything.
(William Shakespeare, As You Like It, Act II, Scene VII)

Shakespeare's words have such power as they tap into some of our elemental fears and desires about life. We age and change, and we rarely have much control over this, yet we still struggle to live well. In Shakespeare's day as in the present, people wish to have a good life and be happy. In modern cultures happiness is everywhere—advertising promising happiness, self-help guides and simple steps to wellbeing and countless books and films depicting the pursuit of happiness. This book emerged from a curiosity about the way that our everyday lives have become framed by this idea of happiness and the desire for a good life. Is the prevalence of ideas about happiness a good or bad thing? Does an increased awareness of wellbeing help people live better lives or create misery as we judge ourselves against the impossibly happy people on TV?

Another reason for my interest in happiness was that many of my friends, now in their 40s, have gone from one crisis to another and seem far less happy than they did 20 years ago. Their jobs and careers had been undermined by insecurity, countless reorganisations, long hours and wages failing to keep pace with the cost of living. Balancing the demands of work with the desire for a home life had also become difficult for my friends as their children arrived, partners worked long hours and elderly family needed caring for too. For some, these pressures were just too much to manage—separation and divorce becoming a feature as we all entered our fourth decade together. My friends had made choices that seemed right at the time but they had been proved wrong—living well and being happy does seem a riddle at times. Could my friends have made different choices or pursued different paths in life to avoid the disappointments of work and the heartbreak of divorce? This is a book then about not just how to encourage positive experiences, which is how happiness is commonly defined, but also the management of the negative events that we all have to endure through life. I understand happiness and the good life in part as the effort we make to have a balance of the good and not so good experiences in life—as too much of either can be problematic. In talking to people about happiness I have been curious about whether we can learn from our experiences and live happier lives. So this book is also about wisdom, and whether an awareness of happiness can help us to develop ways of living that enable us to better balance the good/bad events in our lives. In documenting happiness through the lives of my 19 interviewees I consider whether their stories can offer us some insights into how best to live a good life.

As an academic sociologist I had hoped that I would find some research from my discipline that would help me make sense of these questions about happiness. A quick online search reveals thousands of books about happiness and wellbeing so initially I was optimistic about finding sociological research into happiness. Few of these books, however, are by sociologists for they concern themselves far more with the problems of modern societies than with issues such as flourishing and living a good life. Indeed many sociologists actually see happiness itself as a major problem in modern societies encouraging people to live superficial, consumerist lives that distract us from more fulfilling and socially engaged ways of living. Even where there has been a more concerted study of happiness from psychologists, economists, philosophers and therapists, these approaches tend to offer a partial or idiosyncratic treatment of wellbeing. Psychologists, through employing extensive surveys and experiments, often document the impact of different individual factors such as health or education on wellbeing, neglecting how in reality these different experiences are interwoven in complex ways to influence our happiness. Economists and philosophers also use rigorous methods to generate vast amounts of data on happiness yet use abstract models of individuals, ignoring how people vary culturally and change as they age. Therapy books offer us intriguing personal testimonies about the search for happiness yet focus on personal change as the key to a better life when in fact so much of our wellbeing is shaped by our backgrounds and relationships with others. The aim, therefore, in this book was to research happiness in a way that used the valuable insights of these existing approaches whilst avoiding some of their weaknesses.

Researching Happiness and a Good Life

To explore some of these questions about happiness I interviewed 19 people between 2010 and 2014—some were interviewed just once over several hours but most were interviewed several times. I selected people of different ages, men and women and those from different social backgrounds—some were middle-class professionals and others from less affluent backgrounds. The major aim was to examine what happiness and a good life actually means to these people and whether it varies because of class background, gender or age. I was curious about the way people change over time and how wellbeing varies with age and so I constructed a biographical profile with each person in interviews that charted these temporal developments. I also developed happiness maps for each interviewee that charted how their experiences in different domains such as family, work and education influenced their wellbeing. In particular, I wished to document how people feel about happiness and so I drew on positive psychology to probe how wellbeing is very much a personal and emotional feature of living and something that we reflect on and think through. Yet at the same time I also wanted to move beyond these sorts of psychological questions to also explore how happiness is a much more social phenomenon than just a characteristic of individuals. Here I draw on more sociological thinking to investigate how my interviewees were connected to social networks that influenced their wellbeing. This is an approach that views happiness as something that can be shared, negotiated and struggled over and reflects the tensions and conflicts that emerge out of the unequal power relationships that structure our lives.

For those wishing to explore better ways of living and for researchers curious about happiness it can often seem paradoxical or a sort of riddle. Happiness is a powerful subjective feature of life and we work at trying to be happier, yet these efforts and emotions are influenced by a complex array of relationships and processes we struggle to understand and manage. As my interviewees' stories show, they were at times unsure about how best to enhance their wellbeing as the actions of others and past events had constrained their autonomy. At the same time many also reflected on their lives and developed new ways of living that had helped them to flourish.

Happiness also seems a riddle as different disciplines have a variety of views on how to define happiness, the sources of wellbeing and how best to study these phenomena. Hence this book borrows a little from each to develop a cross-disciplinary approach to wellbeing. From philosophers such as Aristotle who suggested that wellbeing is something we work at in practical ways each day. From psychologists who suggest that happiness involves cognition and the balancing of positive and negative emotions. From sociologists who identify the ways in which shifting economics, cultures and policies structure lives creating patterns of wellbeing. And from Psychoanalytical theory that shows how our deeper selves are made and influences wellbeing in unpredictable ways.

Happiness is puzzling, as the study of an individual's wellbeing has to also consider the interrelationships that individuals have with others. Classical philosophers acknowledged the social nature of happiness, yet much wellbeing research by economists, psychologists and others use experimental models and quantitative surveys that bracket out these social aspects of happiness. As Skidelsky and Skidelsky (2012) have discussed, many philosophers, from Aristotle (2009), Marx (1983), Maslow (2013), Rawls (2005) and Nussbaum and Sen (1993) have developed lists of universal needs or goods that are essential for a 'good life', such as health, prosperity, trust, respect, reason, natural resources and so on. Philosophers then debate how to create 'Good Societies' that enable their citizens to access these goods so they can live well and be happy. I examine how important some of these goods are to the happiness of my interviewees at different points in their lives such as when young, through marriage, mid-life and into old age. But I do this in a way that allows us to see some of the messy interrelationships between these different goods, the negotiations and trade-offs. For the problems with these philosophical debates on wellbeing is how they are far removed from the historically specific, everyday social practice of trying to live well.

When one talks to people about happiness one soon realises that living well and a good life are things we always do with others—yet researchers continue to represent it as a personal characteristic of isolated individuals that can be measured and expressed numerically through equations. This construction of happiness means that it is studied in ways that avoid difficult but essential questions about the way that people live in contemporary societies. We have to acknowledge that one person's happiness can be at the expense of others, which raises questions about the choices people make, how they weigh up the ethics of these decisions and the values that inform them (Sayer 2011). We need to contextualise the success and happiness of individuals by relating their wellbeing to the harms or benefits they create for others. This sort of social analysis of happiness, together with some sort of ethical accounting, is necessary if we wish to develop a critical happiness studies. I have witnessed the 'success and happiness' of many men and women that was only possible because of the hidden sacrifice, loss and disappointments of many partners, children and friends. In these times of austerity since 2008, government policies around the world have impoverished many, whilst social elites, implicated in the crash continue to prosper. How people therefore experience and pursue happiness is interwoven with power and resources and how these are patterned in terms of class, gender, sexualities, 'race' and so on. Hence critical happiness studies links wellbeing to issues of domination, inequality and oppression and by drawing on sociological theories (from Marx and Bourdieu) we can model these connections between people and their struggle to live well and be happy. As C. Wright Mills suggested when talking of the craft of being a sociologist, employing the 'sociological imagination' we have to strive to connect 'personal troubles to public issues' (Mills 1959).We see this interplay of biography and the creativity of people trying to manage wider social processes in the stories that people told me about their happiness. Some were accounts of achievement and good wellbeing, others about luck or misfortune, whilst many more were about coping with injustice and overcoming the many challenges that confront us in our efforts to live well.

Outline of the Book

The first section of the book reviews various literatures from different disciplines that have examined happiness and wellbeing. I offer a short introduction to readers (and my students) of the way that philosophers, psychologists, economists and sociologists have researched happiness. You may wish to skip these chapters if you are more interested in first-hand accounts of how people experience happiness. Nevertheless, these disciplines have provided interesting insights into wellbeing and I employ some of these to develop my own understanding of happiness today—though I also acknowledge there are issues with these accounts. This first half of the book is concluded with a brief discussion of the particular approach I employed in this research—synthesising these different ideas and developing a particular methodology to study happiness empirically. The second half of the book is made up of the empirical chapters that explore the different ways in which happiness was experienced for each of the age cohorts interviewed for the project.

Chapter 2 introduces some of the ideas that philosophers have developed on happiness, in particular around the Greek and Roman notion of happiness as a social process that people work at through their lives. These are ideas that informed how I conceptualised happiness and the way in which it might be studied. I chart the increasingly individualised way that happiness has been understood which reflects processes of modernity and which accounts for why it is often viewed simplistically today as 'subjective, good feeling'. We see the emergence of contemporary pessimism over modernity and the possibility of happiness in the writings of Rousseau that he attributes to the way the State and modern cultures have vulgarised happiness. In simplifying what is a complex process, Rousseau (like later sociologists) argues that our abilities to envision a good life and to flourish have been greatly reduced. In contrast to such pessimism, the teachings of Eastern philosophers and religious leaders continue to offer people hope of a good life and happiness by providing sets of principles and ethics to live by.

Chapter 3 explores some of the contributions that economists have made to the study of happiness and in particular how various factors or events are associated with different levels of wellbeing. Although one can understand the desire to measure wellbeing in the effort to inform policy and create happier societies such ambitions are fraught with challenges. Not least that happiness is experienced subjectively, involving layers of meaning that are accumulated over many years out of the myriad of relationships that make up a life. The analytical models and numerical expressions used by economists abstract from much of this complexity and therefore create images of wellbeing that at times seem far removed from how ordinary people actually live their lives. Certainly economists are correct to highlight the continuing significance of income and affluence for wellbeing (and which informs discussions I had with interviewees), yet they also increasingly acknowledge that many other factors are important for happiness. I discuss the shift from materialistic indicators of development such as Gross Domestic Product (GDP) to more complex measures and how these have been driven by newer ideas around Capabilities and Behavioural Economics that reflect a desire to enhance the modelling of wellbeing.

Chapter 4 offers a summary of research in Positive Psychology and explores debates around the sources of happiness and our abilities to improve our wellbeing. Psychologists have popularised many useful concepts such as the 'hedonic treadmill', which I explore with my interviewees—namely how we adapt to situations, or 'take things for granted' and end up restless and dissatisfied with life. Similarly, other research such as that around 'Flow' offers a good way of exploring the positive experiences in life and how leisure, play and work-life balance are important for happiness. As with economics these efforts to measure wellbeing are to be commended (e.g. see insights into the intensity, duration and frequency of experiences) yet, there are drawbacks with the experimental modelling of happiness. Numerical expressions of wellbeing that have been generated by surveys fail to capture the layers of meaning and emotionally charged nature of our efforts to be happy and live well. And so positive psychology can only take us so far in our effort to understand the place of happiness in people's lives.

Chapter 5 on Psychoanalytical Theory continues some of the debates developed in these earlier chapters, particularly around the notions of self-identity, fallibility and the pursuit of happiness. I incorporated psychoanalytical concepts into the project after listening to many life stories that suggested that people internalise some of the emotional energy associated with early life events that re-emerges to influence later life. The writings of Freud and others suggest that we are often conflicted, have faulty memories and insights and so working at being happy can be challenging for many people.

Chapter 6 on Sociological Approaches to Happiness provides an overview of some of the ways that sociologists have understood happiness and wellbeing. It shows that mainstream sociology has been sceptical about researching happiness systematically as it is conceived as a simplistic, positive and subjective phenomenon that distracts attention from more important issues of inequality, oppression and domination. To study happiness sociologically then one has to use traditional approaches and theories selectively, holding onto these interests in power relationships and social processes (and how social divisions such as class, gender and 'race' function) whilst challenging the scepticism sociologists have towards happiness studies.

Chapter 7 on young people is the first of four empirical chapters that comprise the latter half of the book. I use some of the concepts discussed in the first section to analyse the life stories of these young people. For example, sociologists often view happiness as a problem and so I examine whether young people have grown up with unrealistic expectations of happiness whose pursuit leads to disappointments and sadness rather than good wellbeing. Similarly, I examine whether economists' arguments over income and happiness are reflected in the data and also explore issues such as fallibility that psychologists and psychoanalysts have discussed in their work. I try and make sense of one riddle about happiness—how research often suggests there is a 'crisis of youth' with increasing levels of anxiety and depression amongst young people around the world yet many surveys also indicate that a majority of young people are happy most of the time. What might lie behind this 'ebb and flow' of young people's wellbeing?

If we move forward several years we encounter a group of interviewees now facing the next stage of life—settling down and starting families. Chapter 8 explores the lives of five 'Thirty Somethings' and we can see the effects of class and gender processes on the choices and routes they take through relationships, work and leisure. We witness the social nature of happiness as these participants try and build a life with partners sharing ideas about what happiness and a good life can be. For the women this is a struggle involving trade-offs and tensions with partners who expect too much of them—running homes, caring for children and bringing in a wage. The women's wellbeing suffers and we are shown different ways they manage, coping and adapting in their efforts to flourish.

Chapter 9 examines the wellbeing of interviewees in their late 40s and early 50s. They face new challenges of trying to sustain careers and long-term relationships as well as the demands of teenage children and elderly relatives. Some experience a mid-life crisis so it seems that affluence is no guarantee of happiness. Others who have adopted less materialistic lifestyles seem to be happier, though they lack the conventional consumerist trappings of 'successful lives'. What then is a 'successful life'? Some women continue to suffer abuse yet strive to be happy, drawing on networks of support. We witness the impact of good fortune on wellbeing and how there can be 'silver linings' to some experiences—how eventually some positives do emerge from difficult times in life.

The final empirical chapter, chapter 10, explores how four people dealt with the issues of retirement and some of the challenges of old age such as ill health and bereavement. Some were better able to be happy than others developing new interests and friends in old age whilst others had become lonely and disappointed with life. Although affluence aided these life transitions a biographical approach illustrated how earlier experiences and changes to lifestyles were also important for better wellbeing.

The conclusion sets out some of the main themes that have been developed in this book—about how we can research happiness, drawing on different disciplines and offer a critical, biographical approach that shows the structuring of wellbeing and efforts of individuals to creatively pursue a good life. It reflects on what we have learnt from interviewees about how to be happy at different stages of life. There are many different ways to live—and what makes a good life and successful living is contested that reflect the differences in values and ethics of people. There are things however that people can do to improve their own wellbeing—small changes that together can add up to a better, balanced and more rewarding life. Yet the life stories also showed that wellbeing is also influenced by structures of opportunities and resources they inherit from their families. So government policies, social class, 'race' and gender relations are all implicated in people's wellbeing. For the interviewees these social forces often threatened to overwhelm them yet remarkably they usually found ways of surviving these threats, creating different ways to flourish.

Bibliography

Aristotle. (2009). *Nicomachean ethics*. Oxford: Oxford University Press.
Marx, K. (1983). Alienated labour. In E. Kamenka (Ed.), *The portable Karl Marx*. New York: Penguin.
Maslow, A. (2013). *A theory of human motivation*. London: Merchant Press.
Mills, C. W. (1959). *The sociological imagination*. Oxford: Oxford University Press.
Nussbaum, M., & Sen, A. (1993). Introduction. In M. Nussbaum & A. Sen (Eds.), *The quality of life*. Oxford: Clarendon Press.
Rawls, J. (2005). *A theory of justice*. London: Harvard University Press.
Sayer, A. (2011). *Why things matter to people: Social science, values and ethical life*. Cambridge: Cambridge University Press.
Skidelsky, R., & Skidelsky, E. (2012). *How much is enough? The love of money, and the case for the good life*. London: Allen Lane.

2

Philosophy and the History of Happiness

In many ways this is a book about how to live a good life. The interviews with 19 people will show us some of the different ways that happiness is understood and how people try and cope with the challenges of daily life. These may offer us some insights into our own lives and how we can meet some of the challenges that we face every day trying to live well. Key questions I explore are whether people actually learn from their experiences and get better at living a good life or do we keep on making the same old mistakes all our lives and miss out on that elusive contentment? How do we balance our own desires for a good life with those close to us, or our colleagues or neighbours? How important are simple pleasures or is happiness something that one has to work at over time?

I have always been curious about people's lives, which is why I eventually drifted into sociology as a career. Yet these deceptively simple questions about happiness are ones that are now rarely asked by modern sociologists. Much of this book therefore relies on the insights that philosophers and psychologists have given us into the nature of happiness. We examine this work as it can tell us much about the riddle of wellbeing and the challenges we all face seeking out a contented life. These insights from millennia of writing complement the interviewees'

M. Cieslik, *The Happiness Riddle and the Quest for a Good Life*,
DOI 10.1057/978-1-137-31882-4_2

narratives about happiness and together offer the reader a good balance of everyday examples as well as some classic cases from the past. The reader will see that there is much to learn from earlier studies into happiness and how challenges faced by those living thousands of years ago seem eerily familiar to us today. As you read on you will see that how I devised this project owes much to these classic writings—such as how I defined happiness, who I researched and the sorts of questions I asked interviewees. Though, as the research was exploratory and partially inductive, these features were refined during the research process (Sayer 1992; Glazer and Corbin 1998).

As mainstream sociology has mostly neglected happiness research this book is necessarily a first tentative step along the path to a more rigorous sociological approach to wellbeing (though see, Veenhoven 1984; Hyman 2014; Thin 2012; Bartram 2012). The sociology of happiness has much to offer us as the key principle in sociology is to investigate the experiences of individuals in everyday life, yet also understand how these events and accounts rely on relationships with other people. As C. Wright Mills noted, this problem of humans where our identities and experiences seem very personal yet are structured by power relationships creates a methodological approach that distinguishes sociology from psychology, philosophy and economics. For example, our reliance on others (for love, companionship, employment and so on) means that there are always limits to what we can know about events as they unfold during the day so we can only ever have partial knowledge of situations and incomplete insight into how best to make choices. This fallibility is important in understanding how we sometimes stumble through life and struggle to be happy (Gilbert 2006). As we are inherently social and paradoxically need others to express our individuality, sociology also investigates how individuals are constrained or empowered by our social relationships. So, for example, we acquire ideas (or values) about what is important from those we live with and over time these values can shape how we think and what we do. These values however are never neutral but always reflect their histories and can favour the interests of some people over others. Hence we grow up in Western societies with dominant values that suggest that individual success and affluence are routes to happiness, whereas in Eastern and perhaps some Scandinavian societies there is a greater focus on spiritual wellbeing and the advancement of families and communities (Lu 2001). One consequence of this long-term historical process of sociocultural change is that we rightly or wrongly in the West view happiness in terms of our ability to consume and to choose particular lifestyles. Sociologists then are critical people, keen to ask questions about who gains and loses in life, where does power lie and how do people and societies change over time. It is this critical curiosity developed by sociology that I use to explore how happiness is experienced today by a small group of different individuals I researched over several years.

The Buddha and Eastern Notions of Happiness

In our quest to understand happiness a good place to start is with some of the first thinkers in history to ponder the notion of a good life. The story of the Buddha (Prince Siddhartha Gautama) from the fourth century BC offers a powerful and influential set of ideas about what it is to be human and how we struggle in life to live well. The example of the Buddha provides a template for those who wish to live a more enlightened life telling us a story of how one encounters much pain and suffering and this is an inevitable feature of being human. Unlike modern self-help guides there is no suggestion that we can somehow avoid loss and pain. Neither are we told that a good life is one where we simply work to minimise suffering. Instead for Buddhists the first step towards living a better life is the acceptance of the inevitability of suffering and how we can work (creatively and 'hopefully') to prevent suffering dominating our lives. Next we are told that suffering has its cause or emerges from something—'the Arising' (Schooch 2007: 99). In Buddhism it is the power of our desires that shapes our suffering and so a better life can be achieved by coming to understand our desires and how they drive us in life. In modern times we see this everywhere, such as our pursuit of success at work, our fixation on targets and outcomes or popularity or celebrity, yet we seldom interrogate why we hold these goals and ambitions and where they come from. A happier life can come from developing a way of living that allows you to be more detached from these desires (so-called, Cessation) and devote more time to our connections with others and nature. We see these ideas popularised in contemporary guides on mindfulness (Williams and Penman 2011) that document our obsession with rationalising our lives and reducing them to crude means and ends schema when we should be more mindful of simple everyday experiences—'living in the now'. In the latter empirical chapters we do see some interviewees who come to recognise the power of these insights and change their lives accordingly. And finally there is the 'Path' in Buddhist thinking which conveys the idea that once we begin to reframe our life then we have to begin to change and introduce new daily routines to ensure that we continue to flourish. There are various aspects to this journey of personal change such as meditation (working on the body and mind), cultivating wisdom as well as developing an awareness of morality (the needs of others as well as one's own needs). As Schooch (2007) writes, these Buddhist ideas are as useful today as they were millennia ago for they offer a much more personal approach to living than established religions as there are no idols or Gods to worship. Yet Buddhism still addresses some of the fundamental questions about how to live a good life. To reduce complex processes (such as learning or health care) to simple measures (such as grades or outcomes); to obsess about to-do lists and goals may all seem very modern afflictions but we can see they are problems with ancient origins. Buddhism's depiction of life's challenges seem particularly prescient today—trying to be compassionate whilst satisfying our own needs, and trying to hold onto the truly important things in life such as relationships with others and with nature. These were struggles that all of my interviewees faced at various points in their lives and they developed their own ways of managing these tensions—some more successfully than others. Just as Buddhism has been influential today so it was influential in shaping the views of other philosophies on the good life as we now see with the ancient Greeks who also grappled with some of these fundamental questions of how to live a good life.

Classical Greek Writings on Happiness

Happiness researchers are often asked, 'what is the secret of happiness', by which people usually mean, 'what do I have to do in order to feel more positive about my life'. Many want a nice simple response so I usually offer the bland reply that a bit more money can cheer people up but ultimately contentment comes from the quality of the relationships you have with others. Within a few minutes and a bit more thought we usually agree that what appears quite simple at first raises a host of questions about the relative contribution of things and people to ones wellbeing. This curiosity about happiness and how to achieve it is certainly not new as it was a feature of debates in classical societies two thousand years ago. Reviewing these classical works is useful as it offers us insights into how we can come to define what we mean by happiness and how it might be achieved, for in classical times as now these were hotly contested issues.

As with Eastern philosophy, early Greek classical thinkers such as Homer and Herodotus believed that the desire for pleasure and the pursuit of a good, contented life was a universal feature of human beings (McMahon 2006: 3). All of us then have the ability to experience these 'good spirits' or 'Eudaimon', though some are more fortunate than others in how much of this spirit they enjoy, for these thinkers believed that the degree of happiness one encounters in life is shaped more by luck or fortune or fate—the acts of gods rather than our own endeavours. The Greek tragedy plays, like the later Shakespeare provide some of the classic accounts of how misfortune or good luck can change lives and informs our views on happiness that have shaped Western culture. As we see in the following chapters all of my interviewees spoke of the role of good and bad luck in shaping their happiness and this raises interesting questions about how much control we actually have over our abilities to be happy.

In their writings we see early Greek writers such as Socrates and Plato grappling with this conundrum of how to define happiness. Whether it involves good feeling and fleeting, positive sensations or more enduring features that today we would suggest are embodied, residing in personality, a way of life or habits and routines. Both Socrates and Plato recognised the distinction between simpler, everyday pleasures—Hedonia or hedonism—and the longer-term process of living a contented life that cultivates our inner good spirits or Eudaimonia (McMahon 2006: 36–40). They develop as Buddhism did the distinction between more superficial forms of happiness and more enlightened enduring wellbeing and that the good life should have some form of balance of these different sorts of happiness. This issue of balance in life proved significant for many interviewees trying to enhance their wellbeing, trying to make choices between different activities and their contribution to quality of life. These everyday tasks of balancing wellbeing involve judgements of value between different varieties of happiness and this can be challenging. Indeed philosophers have debated these issues for millennia—a problem that has continued to the present day where simple everyday pleasures (such as listening to pop music) can be derided for being hedonistic whilst more enduring activities (that take time to cultivate, such as playing an instrument) that generate contentment are accorded higher worth (see Sayer 2011).These disputes and dilemmas featured in accounts from interviewees and illustrate some of the challenges people faced navigating their way through life in an effort to flourish. A key question we all face is whether the ability to live well is down to fate, luck, skill or the choices we make between simpler pleasures or more enduring contentment.

Aristotle famously tried to resolve these challenges of happiness in his book, *Nichomachean Ethics* (2009), popularising the image of 'the craftsman of life' and Eudaimonic happiness. If humans can employ their powers of reason (to reflect on life and make informed choices) then making the right decisions in life will help us to be good or virtuous and with it bring a lasting enduring happiness. Reasoning in everyday life should also allow us to develop 'a rule of thumb' approach to decisions and choices and so help us be moderate and avoid extremes—the so-called doctrine of the mean (McMahon 2006: 48) which can promote a more enduring happiness. These are ideas that have much resonance today with our contemporary concerns for work-life balance. My interviewees spent much time trying to balance the many demands on their time and for some it was a major source of stress in their lives and a key barrier to a happier life.

As we saw earlier with Buddhist teachings Aristotle's notions of moderation and the craftsman conjure up the image of the individual working each day, taking considered choices which slowly over time help shape a balanced, contented and harmonious life. There is very much an emphasis here on practical social activity, of learning (from others and past mistakes) and developing skills for living that echo Buddhist's teachings on exercising a vigilant, mindful and disciplined approach to life. Nevertheless, despite this emphasis on happiness as a process of practical reasoning, Aristotle also acknowledged particularly in his later works such as Rhetoric the other elements of happiness that many value today such as the role of fortune and 'external goods' such as wealth, family, friends and internal goods such as a good mind and physical health; factors that subsequently have been the subject of much investigation by economists and psychologists.

The significance of these classical thinkers is they offer us the first multidimensional conception of happiness, acknowledging how we employ our reason to pursue our desire for the good life yet are constrained by luck or fate and the actions of others. This discussion of Eudaimon or the journey of happiness pushes us towards an understanding of happiness that is much more of a social and personal process than the one offered by many contemporary commentators. Many researchers today and particularly positive psychologists and economists tend to freeze these processes preferring models of individual happiness, lists of individual characteristics, factors and causes and effects. Whereas in this book I have tried to capture this more social and situated process of happiness that is context-dependent, exploring how interviewees worked each day at their wellbeing, thinking through their interests, struggling to do the right thing in the face of a myriad of constraints and competing demands from others. The problem with much positivistic research into happiness as we see with contemporary survey–based approaches is the focus on the isolated or abstract individual and their happiness which neglects the daily ethical challenges and unequal power relationships people face navigating their social relationships whilst pursuing the good life. What sort of insights into happiness do we generate if we only concern ourselves with snapshots of individuals when for example one person's positive wellbeing may be at the expense of others—behind the successful academic or politician and their smiling faces may lie a long tail of disappointed and miserable children, partners and colleagues. As a way of combating some of these weaknesses of existing research we need to shift attention from the individual onto those who are near to them and then back again onto the subject of analysis, in the process analysing the trade-offs and the winners and losers as people make ethical choices and pursue their lives and happiness (Coulthard and Brittan 2015).

Roman and Christian Writings on Happiness

Later Roman and Christian writers were greatly influenced by Greek philosophers. In particular they acknowledged the role of wider social change, such as how shifting political systems and changing values can influence personal wellbeing. This has become a major problematic in happiness research and sociology more generally centred around the way that social processes frame our experiences, life chances and identities. These writings have particular relevance today as we live through a time of economic restructuring, austerity and the insecurities and inequalities these have brought. The works of Epicurus and Zeno, the father of Stoicism (and followers Diogenes and Epictetus) mark a turning point in Athenian philosophy as these thinkers wrestled with a declining Athens and its consequences for the wellbeing of its citizens (McMahon 2006: 53–59). We see an increased emphasis on the responsibilities that individuals have for shaping their own wellbeing. The decline of the Athenian state was causing great anxieties about the ability of the state to ensure the welfare of citizens shifting responsibilities to individuals to manage these insecurities as best they could. There was a recognition that although developing an informed craftsman approach to life promoted personal happiness, wider social developments can easily threaten the wellbeing of individuals (Schooch 2007: 173). For the modern reader these concerns of Greek philosophers seem eerily familiar—as we see welfare reforms in Britain creating new hardships whilst government ministers exhort citizens to become self-reliant and resilient (Duncan Smith 2015).

During this period of decline in the Athenian society the Greek Stoic writers promoted ascetic doctrines as a route to happiness. These encouraged individuals to use reason and self-control as a way of managing the insecurity of the times. These philosophies found favour in Roman times influencing the work of later Stoics such as Seneca and Cicero and Epicurans such as Lucretius for there were parallels between the development and decline of Athens and that of the Roman Empire (McMahon 2006: 53). The excesses of Rome in particular led these later philosophers to question the merits of civilisation contrasting city life with the simpler and more natural experiences and happiness of the countryside—a dualism that has proven timeless. The writings of Horace and Virgil at this time extolled the virtues of plain living and expressed in popular sentiments such as 'Carpe Diem', or 'seize the day'. These writers established important debates that have shaped happiness studies as we see with these tensions between city life and living close to nature. Many writers as diverse as Marx (1984) and Csikszentmihalyi (2002) have noted the role of nature in flourishing and indeed my interviewees spoke of the restorative powers of nature such as a simple walk through a wood or along a beach. For many, happiness could be had by escaping urban living and finding joy in nature, even gardening on ones allotment.

The latter stages of the Roman Empire saw the rise of Christian theology and new thinking on happiness and the good life. Although these are ancient ideas, they are relevant for my interviewees who were trying to live well. The Christian scriptures popularised many of the ideas on happiness that we saw in earlier Greek and Roman writings, such as the inevitability of suffering, the need for self-control, distinctions between simpler and enduring wellbeing, the use of reason and an ascetic sensibility to help in one's journey through life. The writings of St. Augustine and Thomas Aquinas illustrate some of these elements of Christian teachings and how they can be seen as guides for happiness and the good life. In many ways these early philosophical writings were the first examples of self-help guides and the biblical stories and the commandments in particular were used by ordinary people to make sense of their suffering and to promote better wellbeing. St. Augustine and Thomas Aquinas like later Christian figures popularised key tenets of biblical teachings, such as the role of sacrifice and good works in the pursuit of virtue and an effort to get closer to God. Here we see notions of compassion, forgiveness and altruism playing a role in how individuals should over time cultivate a pious sensibility and in so doing would enhance their wellbeing through salvation in the afterlife. As we go on to explore through the lives of interviewees these aspects were important features of the stories that respondents recalled about their wellbeing and how happiness functioned in their lives. Many spoke about the significance of having some sort of framework of beliefs and values that could help make sense of life as well as sets of relationships, activities and rituals that allowed for these shared ideas to be employed. For some this still operated in ways that would be familiar to early Christians via religious stories and ceremonies

...... ւnese had become secularised but the storytelling, the shared beliefs, friendship and rituals still remained and were significant for good wellbeing.

Happiness and the Rise of Modern Societies

From the 1400s there was a growing interest in happiness, as during this period there was a gradual critique of the power of religion and monarchy which led thinkers to frame happiness debates more in terms of the individual seeking happiness than in the role of divine influence or fate on wellbeing. This questioning of older traditions and the beginnings of individualism was seen in events such as the Reformation and Luther's radical claims that believers can have a personal relationship with God rather than faith being mediated by priests and the church hierarchy. Such changes ushered in new ideas about happiness and religion, shifting the emphasis from wellbeing emerging from strict adherence to doctrines to one where individuals have more freedom to pursue their faith and happiness in their own way. The significance of this today is how we expect to be happy—there is a 'right' to have a good life, but that such freedoms also bring a responsibility for us to work to achieve this happiness. If people fail in this pursuit of happiness then these ideas and values can also demonise the unhappy. We see here the beginnings of a modern understanding of 'happiness as a problem', one that is used to promote a pessimism about modernity.

As McMahon writes (2006), the Civil War in England, the French Revolution and the Independence of America were also key events that helped forge our modern ideas of happiness. These events for many reflected the shifts in thinking of the time, particularly around the rights and freedoms of individuals, the emergence of nation states and who controls them which in turn gave these debates new impetus. The questioning of traditional institutions such as the monarchy and emergence of republican ideas posed questions about the power of humans to create and manage societies for themselves. It led to debates about what it meant to be a citizen and the relative responsibilities of individuals and State to ensure the wellbeing of citizens. We see an example of this construction of happiness as part of these wider social and political developments in the American Declaration of Independence.

> We hold these truths to be self-evident: that all men are created equal; that they are endowed by their creator with unalienable rights; that among them are life, liberty and the pursuit of happiness. (Jefferson 1776)

Many writers during this period laboured over the tensions in these radical ideas as we see with Bentham's concern for modern society to aid everyone to be happy yet in a way that does not undermine the common good (Bentham 1776). Such concerns are just as pressing today for how should the State support individual wellbeing? How can we develop a more relational auditing of wellbeing so we can analyse how the pursuit of happiness for some does not create suffering for others?

The shift in power from the church and monarchy towards individuals led many thinkers to ponder the challenges of these new responsibilities of citizens to live good and happy lives. We have then an emerging pessimism about the ability to be happy that characterises much modern thinking about happiness, so that happiness has paradoxically become a problem in modern times (Cieslik 2015). Locke for example wrote of the inevitable restlessness of humans and their inability to be content as modern societies create new desires and wants (McMahon 2006: 180–182). Mill, echoing the writings of the Classic Greeks, wrote of the curse of reason, that to fixate on being happy is no route to happiness for it emerges as a by-product of living (McMahon 2006: 347–348). We see the development of the very modern dilemma of happiness created by these shifting power relationships, namely that greater numbers of individuals have new freedoms to think about and pursue a good life but with them come increased expectations and possibilities for disappointment. Strangely, as modern societies emerge so the shifting ideas around happiness create new risks of increased suffering and unhappiness. Hence many modern philosophers are sceptical about the merits of happiness research for they believe that modernity has been the harbinger of suffering (Nietzsche 2003; Wilson 2008). Many contemporary sociologists critical of happiness research employ this line of argument. Namely, how large corporations promote unrealistic expectations of the good life through endless consumerism and our inevitable disappointments then encourage yet further consumerism (Davies 2015).

Some commentators suggest that the quality of life of most individuals in Western societies have improved with industrialisation (Veenhoven and Vergunst 2013). Advancements in food production, the expansion of scientific understanding and medicine as well as growing prosperity were all significant for enhancing wellbeing. The data on morbidity and mortality all suggest that from the 1500s to 1800s life became better for the majority of individuals. Nevertheless, from the 1700s many writers came to question the ideas of social progress that had become influential since the Renaissance. Here we see an early example of this riddle of happiness. How growing prosperity and improvements in the quality of life are accompanied by pessimism and doubts about wellbeing illustrating a disconnect between shifting objective conditions for people and their subjective assessments of their lives. Rousseau posed many of the critical questions that would come to characterise later sociology, particularly around the way that modern societies appear to promote wellbeing but instead create false needs and superficial ways of living—an egoism or self-love that corrodes the true social and creative nature of humans. The very modern message here is that the use of reason, new forms of sociability, scientific understanding and the new forms of state organisation are no guarantee of authentic happiness. Indeed the new freedoms of modernity create new forms of critical consciousness that are also the sources of misery and suffering.

> We daily deplore human miseries, and we find the burden of existence rather hard to bear, with all the ills that weigh it down. Always asking others what we are and never daring to ask ourselves…In the midst of so much philosophy, humanity, politeness, and sublime maxims we have merely a deceitful and frivolous exterior: honour without virtue, reason without wisdom and pleasure without happiness. (Rousseau quoted in McMahon 2006: 238)

This pessimism takes on a more profound form in the work of Nietzsche who holds out little optimism for the wellbeing of individuals in modern societies. He suggests that scientific and religious doctrines only ever offer partial ways of making sense of life. These beliefs only offer so much meaning to the individual, can only ever answer so many questions about how to live and so suffering is inevitable as modern individuals come to recognise the inevitable loneliness and powerlessness that flows from being human.

This brief overview of philosophical writings has shown how the ideas around what happiness is and how it can be achieved have evolved with wider social changes. These insights are useful in our effort to explore how happiness functions in the lives of ordinary people today. The sorts of questions I asked my interviews reflect these historic debates and inform the later chapters of this book. How much is happiness about everyday pleasures or is it about a more profound deeper sense of contentment? What role does luck and fortune play in people's understanding of their wellbeing or do they feel they have a greater responsibility for their happiness than serendipity? If modern cultures have an individualistic conception of happiness and we are bombarded with idealised images of happy lives then does this paradoxically create misery and suffering as people struggle to meet these increasing expectations of a happy life? What of religious or spiritual doctrines—do they still offer meaningful ways of making sense of life and help people to be happy? Despite the emphasis on individualistic notions of wellbeing, does the growing complexity and sociability of modern societies produce forms of social happiness infused with tensions and conflicts? How do we make sense of these divisions and challenges—are the poor less happy than the rich, are women and minorities more subject to sadness and misery than males and the majority? In the next chapter we examine the development of the economics of happiness and its effort to answer some of these questions.

Bibliography

Aristotle. (2009). *Nicomachean ethics*. Oxford: Oxford University Press.

Bartram, D. (2012). Elements of a sociological contribution to happiness studies. *Sociology Compass, 6*(8), 644–656.

Bentham, J. (1776). *'Preface to the first edition', A fragment on government*. In Mack, M. P. (Ed.). (1969). *A Bentham reader*. New York: Pegasus Press, p. 45. In McMahon, D. (2006). *Happiness: A history*. New York: Grove Press, p. 212.

Cieslik, M. (2015). Not Smiling but Frowning: Sociology and the Problem of Happiness. *Sociology, 49*(3), 422–37.

Coulthard, S., & Brittan, E. (2015). Waving or drowning: An exploration of adaptive strategies amongst fishing households and implications for wellbeing outcomes. *Sociologia Ruralis, 55*(3), 275–290.

Csikszentmihalyi, M. (2002). *Flow: The classic work on how to achieve happiness.* London: Harper and Row.

Davies, W. (2015). *The happiness industry: How the government and big business sold us wellbeing.* London: Verso.

Duncan Smith, I. (2015). *Work, health and disability.* Speech at 'Reform', Great Peter Street, London, 24th August. Retrieved from http://www.reform.uk/publication/rt-hon-iain-duncan-smith-mp-speech-on-work-health-and-disability/

Gilbert, D. (2006). *Stumbling on happiness.* London: Harper Perennial.

Glazer, B. G., & Corbin, J. (1998). *Basics of qualitative research: Grounded theory procedures and techniques* (2nd ed.). London: Sage.

Hyman, L. (2014). *Happiness: Understandings, narratives and discourses.* London: Palgrave.

Jefferson, T. (1776). *Declaration of independence.* Retrieved from http://www.archives.gov/exhibits/charters/declaration_transcript.html

Lu, L. (2001). Understanding happiness: Chinese folk psychology. *Journal of Happiness Studies, 2*, 407–432.

Marx, K. (1984). *Marx: Early writings.* Harmondsworth: Penguin.

Nietzsche, W. F. (2003). *The genealogy of morals.* New York: Dover Publications.

Sayer, A. (1992). *Realism in social science: A realist approach* (2nd ed.). London: Routledge.

Sayer, A. (2011). *Why things matter to people: Social science, values and ethical life.* Cambridge: Cambridge University Press.

Schooch, R. (2007). *The secrets of happiness: Three thousand years of searching for the good life.* London: Profile Books.

Veenhoven, R. (1984). *The conditions of happiness.* Lancaster: Kluwer Academic Publishers.

Veenhoven, R., & Vergunst, F. (2013). *The Easterlin illusion: Economic growth does go with greater happiness.* EHERO Working Paper, Erasmus Happiness Economics Research Organisation, 2013/1, 23. Retrieved from https://mpra.ub.uni-muenchen.de/43983/

Williams, M., & Penman, D. (2011). *Mindfulness: A practical guide to finding peace in a frantic world.* London: Piatkus.

Wilson, E. (2008). *Against happiness: In praise of melancholy.* New York: Sarah Crichton Books.

3

The Economics of Happiness

[T]here is another greater task, it is to confront the poverty of satisfaction—purpose and dignity—that afflicts us all. Too much and for too long, we seemed to have surrendered personal excellence and community values in the mere accumulation of material things. Our Gross National Product, now, is over $800 billion dollars a year, but that Gross National Product—if we judge the United States of America by that—that Gross National Product counts air pollution and cigarette advertising, and ambulances to clear our highways of carnage. It counts special locks for our doors and the jails for the people who break them. It counts the destruction of the redwood and the loss of our natural wonder in chaotic sprawl. It counts napalm and counts nuclear warheads and armored cars for the police to fight the riots in our cities. It counts Whitman's rifle and Speck's knife, and the television programs which glorify violence in order to sell toys to our children. Yet the gross national product does not allow for the health of our children, the quality of their education or the joy of their play. It does not include the beauty of our poetry or the strength of our marriages, the intelligence of our public debate or the integrity of our public officials. It measures neither our wit nor our courage, neither our wisdom nor our learning, neither our compassion nor our devotion to our country, it measures everything in short, except that which makes life worthwhile. Robert F. Kennedy (1968)

M. Cieslik, *The Happiness Riddle and the Quest for a Good Life*,
DOI 10.1057/978-1-137-31882-4_3

This chapter explores some of the ways that economists have studied happiness. In its modern form since the Western Enlightenment, economics has drawn on mathematics to model factors linked to wellbeing as a way of assessing social progress. The growth in affluence, prosperity or income has often been seen as a proxy for good wellbeing—the argument being that material resources offer the freedoms in Western societies to improve the quality of our lives. As I go on to show this interest in material resources for good wellbeing has characterised much of the debate in economics. This is hardly surprising given the values and ideas that underpin Western societies. As I discuss elsewhere in this book, Max Weber (1990) wrote about the Protestant Ethic—how increasing materialism emerged from Christian beliefs promoting the notion that happiness could be derived from affluence and prosperity. Yet since the latter half of the twentieth century there has been much debate over the weaknesses of materialistic indicators of social progress and wellbeing and a call for a more sophisticated understanding and measurement of 'a good life' (OECD 2004). Economists have been torn between the desire for relatively simple measures of wellbeing (that allow for macro-level research and analysis) and the recognition that happiness and a good life are far more complex than the proxies they employ in their models. This riddle of happiness has bedevilled economics and the policy-making and political debates that have sprung from it. Robert Kennedy expressed his dismay at this state of affairs in 1968, speaking of how modern societies use crude accounting measures that purport to assess quality of life when in reality they abstract from the very things that make such life worth living. In this chapter I illustrate how economists have tried to grapple with these dilemmas, showing that much of their work has offered considerable insight into the many different factors that influence wellbeing. As I illustrate in the later empirical chapters in this book, the interviewees acknowledged the significance of factors such as marriage, good health and values to their wellbeing—all of which have been extensively researched by economists. Yet at the same time they also noted that their happiness was framed by subtle variations of interpretation and meaning that arose from their biographies, their beliefs, values and social relationships over many years—features that economic models often struggle to apprehend. Hence I argue in this book that the study of happiness requires a synthesis of techniques from different disciplines, marrying the insights of economics with those of psychology and sociology. Although numerical expressions of wellbeing can offer insights into how we live at the same time we also need some narrative accounts that illuminate how happiness is something that we strive for through our lives and is open to interpretation and contestation.

The Emergence of the Economics of Happiness

As we saw in previous chapters, the conventional modern history of happiness often begins with the work of Bentham (1776) and he is also a founding figure in economics. He was one of the first modern thinkers to assess the effect of different factors on the wellbeing of individuals using his so-called felicific calculus. Bentham believed that all individuals wished to pursue happiness and sought to calculate how different pleasures and pains contributed to an overall measure of individual wellbeing and in turn an estimate of happiness for wider society. Bentham like contemporary economists wished to discover which factors led to the greatest pleasures or pains so to help engineer progress to better societies—'to judge the utility of things'. This endeavour was a reflection of the social changes of the time as we saw in the French and American Revolutions—the belief in government and social policies to bring about positive social reform. And so Bentham was concerned to assess the utility of different laws, income and various freedoms for happiness. Yet as McMahon documents (2006: 219), although this utilitarian approach has been influential in focusing attention on the effects of things on wellbeing, Bentham's efforts to adequately measure pleasure or pain were problematic. He spoke of the duration, certainty, proximity or intensity of positive and negative experiences but struggled to assign different values to these many different events. Bentham came to realise that subjectivity introduces endless variability to how different factors may be experienced by different people. In an undated manuscript Bentham conveys his sense of frustration at the challenges posed by the study of happiness,

> 'Tis in vain, to talk of adding quantities which after the addition will continue distinct as they were before, one man's happiness will never be another man's happiness: a gain to one man will not be a gain to another: you might as well pretend to add 20 apples to 20 pears. (Bentham 1989, in McMahon 2006: 99)

Instead of incorporating the complexity of subjectivity (or indeed society) into their research, economists since Bentham have used idealised models that assume people are well informed about market conditions and make rational choices or preferences in pursuit of their self-interest (Frey and Stutzer 2014). As part of this approach they usually employ monetary or financial expressions of people's desires and needs. Hence the value or utility of things (and their influence on wellbeing) can be judged by how people allocate their financial resources to different goods and services. Many economists since Bentham's time, therefore, have explored the relative worth of goods and services that are seen as proxy measures of their impact on personal and social wellbeing (Frey and Stutzer 2002). In the twentieth century this logic has translated into societal measures of Gross National Product (GNP) or Gross Domestic Product (GDP) that were adopted by the International Monetary Fund and World Bank in the late 1940s. These allow us to assess changes to economic growth and wider societal prosperity by adding up the monetary worth of goods produced each year. For many decades such measures of GDP/GNP were viewed as indicators of social progress and with it crude proxies for wellbeing employed by politicians and policy-makers alike (Fox 2012).

Some Problems with the Economic Modelling of Happiness

Economics has been influential in shaping the nature of happiness studies with their mathematical modelling and idealised scenarios being adopted by other disciplines, such as psychology and the social sciences more generally. They usually employ Likert scales and Cantril's ladder (a scale in the form of a ladder) in their surveys to explore self-reports of life satisfaction or affect (positive or negative emotions such as joy or anxiety)

(Cantril 1965). Yet, for sociologists this approach, in an echo of Robert Kennedy's concerns, tends to abstract from many of the features of life that give it its richness and uniqueness. From the founding sociologists onwards there has been an interest in how actors are not always rational or well informed about the choices they make in life. The cultures we acquire and the power relationships that structure our lives all shape our ideas and in turn the choices we make—yet economist's models bracket out these sources of complexity and human diversity. Happiness and notions of a good life are also emotionally charged with complex layers of meaning that people accumulate over many years—how can simple survey questions really capture this complex reality? For example, wellbeing surveys developed by economists will ask very different people the same question about how satisfied they are with their life. As I discuss later on in the empirical chapters, an old and a young person may both respond that they are not very satisfied with life (answer 2 on an 11-point scale on life satisfaction) yet what this dissatisfaction actually refers to empirically may be very different for each. The older person may have poor health, have lost a loved partner and face an uncertain future, whilst for the adolescent their dissatisfaction may stem from worries about school exams and arguments with peers. Can we really claim that both people have similar (dis)satisfaction with life when their interpretations of these questions, rooted in very different life experiences, are so widely different?

Sociologists also acknowledge in their work that people by their very nature are social animals and need others and so research methods should try and explore how something like happiness emerges out of these social dynamics. Hence the problem with large-scale wellbeing surveys is the assumption that happiness or wellbeing exists as a characteristic of an individual. This sort of ontological position sees people as isolated individuals connected by crude mechanical relationships. This greatly simplifies the messy, contradictory ways that negotiation, adaptation and struggle are woven into the fabric of the many different forms of association that people experience and their influences on wellbeing. For example, the anthropologist Neil Thin (2012) writes of the place of love in wellbeing but how can surveys begin to do justice to the machinations of love? The people I interviewed for this book all spoke about the central role that love has in their lives—being loved as a child, falling in love or the love a parent has for a child. The way that we care for others and need caring for in different ways as we age and how emotions are attached to these experiences are at the heart of wellbeing (Smart 2007: 59) yet surveys struggle to make sense of these features of life. As this economic modelling of wellbeing is far removed from how we actually live and how happiness is experienced, such surveys for sociologists at least can only offer partial insights into the nature of happiness. If we are curious about happiness and how it functions in people's lives we have to move beyond these types of economic models and surveys combining this data with other research techniques employed by sociologists.

During the 1970s economics debated their analytical models and the significance of income for happiness. These disputes illustrated how contentious happiness research can be and how wellbeing is inherently political and ideological. For governments and elites would like ordinary people to believe that economic growth leads to prosperity and happiness. If they do not then there is cause to question the political and economic status quo. For my interviewees the significance of these issues was that they had all grown up believing that hard work and prosperity can lead to happiness. If this is not the case, then how does one live? Cross-national research into life satisfaction by Richard Easterlin (1974) posed many of these questions, causing much controversy. He suggested that though richer people tend to be happier than poorer ones living in the same society, as societies became more affluent the average levels of life satisfaction do not necessarily rise with increases in average income as might be expected. This research received so much attention as it questioned orthodox economics and one of the pillars of capitalist societies—that increasing income or affluence is the route to happiness. The policy implications for politicians were that they should reassess their fixation on economic growth (GDP) as a key political goal for the wellbeing of citizens is not simply linked to growing prosperity. For my purpose here, one of the insights from this debate is the question about the role of income for people's wellbeing. Throughout this book I document how the happiness of different interviewees (who varied by age, class and gender) were affected by changes in prosperity. As interviewees became more affluent did they become happier? If the incomes of participants declined did this undermine their wellbeing? Or as Easterlin suggests are the links between income and happiness more complex than this—are some people able to isolate their wellbeing from changes in income as they are adaptable or resilient?

Such was the controversy created by Easterlin's original work that economists are still exploring his claim that richer societies do not necessarily become happier ones (see also Easterbrook 2003; Lane 2000; Schwartz 2004). Numerous economists have subjected Easterlin's original data to re-analysis or conducted similar research with new datasets to explain why rising incomes did not lead to increasing life satisfaction (Blanchflower and Oswald 2004; Graham and Pettinato 2002). The different explanations for this anomaly have become important areas of debate in happiness studies and inform some of the interviews I conducted. Stevenson and Wolfers (2008) summarise these issues around the so-called Easterlin Paradox suggesting that one explanation for slow growth in happiness rates in the USA despite growing prosperity could be attributed to social inequality—only some citizens actually experienced growing incomes. They also noted how people make judgements that influence their wellbeing through comparison with other people and hence even if incomes rise then happiness levels may be depressed if people see others doing well. Paradoxically, economists tell us, that the affluence of others can have a corrosive effect on the wellbeing of the poor even if the incomes of the poor are rising. Sociologists have also noted this significance of reference groups for people's wellbeing. Veblen, for example, coined the term 'conspicuous consumption' to denote the way our consumption choices (and the satisfaction derived from them) are often informed by our perceptions of others. We often purchase things then to impress other people as much as to satisfy our own needs. Bourdieu (2010) researched the role of recognition that we need from others when pursuing particular lifestyles—noting the way that prestige or status (forms of distinction) can flow from everyday displays. This role of reference groups in relation to happiness and people's lifestyle choices was a topic I was keen to investigate in the interviews I conducted. In particular, whether it was possible to resist this urge to pursue a conventional, materialistic life that involves 'keeping up with the Jones' and develop alternative ways to flourish at a time when austerity in Britain had threatened the incomes of many people.

Finally, a key finding by Richard Easterlin was the claim that in poorer societies people became happier as the wider society became richer but that once per capita income rises to around $15,000 then these increases in happiness slow down. The notion of a satiation point was suggested to account for this finding, whereby individuals adapt to increases in income and further income growth has less impact on wellbeing. This research led to further investigation into what has become known as 'habituation' or the 'hedonic treadmill' effect (Layard 2005: 48). The idea behind these concepts is that humans adapt to their conditions, seeking out new challenges and experiences and so increasing income and the things that it can buy do not necessarily lead to enhanced wellbeing. An empirical question here is whether people come to realise that the good feelings associated with purchases wear off quite quickly? Do people wise up to this notion of habituation and the hedonic treadmill and seek out alternative sources of wellbeing—through relationships rather than things?

The Capabilities Approach to Wellbeing

Since Easterlin's famous research in the 1970s, economists have been grappling with the challenges of studying happiness recognised by early economists such as Bentham (1776), Adam Smith (2012) and Hutcheson (1725). Although the pursuit of happiness and a good life appears a simple activity that people do, its analysis requires models that can accommodate the complexity of individuals, their subjectivities and the societies in which they live. Economists have responded by refining their theories and methods as seen with the work of Amartya Sen (1999) and Martha Nussbaum (2000) on capabilities.

Nussbaum and Sen (1993) have promoted the notion of 'Capabilities' where instead of a traditional focus on abstract models and human rights they draw on Marx and Aristotle, to develop far more complex conceptions of the individual and society. Instead of reducing wellbeing to notions of satisfaction, the capabilities approach expands the understanding of the individual to include a variety of skills and activities necessary to flourish (such as the use of senses, play, emotions, relations with nature)—as we have seen in Marx's writing (1983) on non-alienation or

Aristotle's (2009) discussion in the Nicomachean Ethics. For Sen and Nussbaum a good life involves being able to develop these features of our nature, such as creativity and sociability that are at the heart of living well. Unlike orthodox economics' treatment of wellbeing that models individuals as isolated 'pleasure-seekers' the capabilities approach acknowledges that wellbeing is greatly influenced by wider social relationships and the political and cultural processes in different societies. A good society—one that promotes social justice—involves the State using public policy to ensure that citizens have a range of opportunities to develop their potential or capabilities. These are conceptions of wellbeing that sociologists would easily recognise, for a good life is seen to be dependent on the sorts of social structures we find in a society as well as the complex, interpretative ways individuals engage with these processes in pursuit of wellbeing. Thus when mapping my interviewee's biographies I linked their life events to social processes to explore the interplay of structures and agents (Hockey and James 2003).

The Capabilities approach to wellbeing has been influential since the 1990s, informing efforts to develop more complex instruments to measure and promote wellbeing around the world, such as we see with the United Nations Human Development Index (UNDP 2015) the Kingdom of Bhutan's surveys on Gross National Happiness (Bhutan 2012) and the French Government's report in 2009 into alternatives to GDP (Stiglitz et al. 2009). These measures combine the per capita income data with other indicators such as educational attainment and life expectancy to create indices that allow for the ranking of countries around the world. Countries such as Sweden, Norway and the Netherlands tend to feature at the top of these league tables for Human Development, whereas the UK in comparison does rather less well. The British Government's Office of National Statistics (ONS) also began to measure national wellbeing in 2012 (ONS 2011), its surveys measuring people's responses to questions on life satisfaction, meaning and different aspects of affect. They draw on methodologies developed by economists to show variations in wellbeing over time and correlated against different factors such as age, gender and region. The wellbeing data is generated via the annual population surveys of 165,000 individuals in Great Britain and showed that people's wellbeing in most parts of the UK had improved since 2012, despite austerity welfare policies and the challenges these posed for household incomes (ONS 2015a). The 2015 survey showed that on an 11-point scale for life satisfaction there were some small regional differences in average wellbeing—around 7.50–8.00 with Northern Ireland appearing to produce higher averages and London somewhat lower satisfaction levels. The majority of those surveyed answered in the top three sections of the first three questions (8, 9 or 10 out of 10)—saying that they were mostly satisfied, found meaning and were happy with life. For the final question on anxiety a majority said that there were never/rarely anxious.

Contemporary Economics of Happiness: 'Behavioural Economics'

Recent decades have witnessed an explosion of research by economists into different aspects of wellbeing. Much of this work has tried to offer more complex notions of individual behaviour and social processes than previously. Deaton and colleagues (Deaton 2012; Kahneman and Deaton 2010), for example, have employed more subtle notions of subjective wellbeing than earlier research, distinguishing between satisfaction, having meaning in life and different emotional states (such as joy or anxiety). Such a typology allowed them to explore the Easterlin paradox suggesting that rising income has an influence on levels of satisfaction but less of an impact on the other aspects of wellbeing (Kahneman and Deaton 2010: 16492). Blanchflower and Oswald (2008) have also conducted numerous studies into the determinants of subjective wellbeing; for example, around the way that aging can frame the quality of life of individuals and in particular how middle age can be associated with poor wellbeing. This data, which produces a U-shaped wellbeing curve across the life course (with higher wellbeing levels for the young and older individuals), was a phenomenon I was keen to explore with my respondents. Is a mid-life crisis something that all people, irrespective of social background, experience, or is it restricted to white affluent men who have had relatively straightforward routes through life? What might be the causes of a mid-life crisis—is it due to the changing perceptions that people have of their lives which account for their disenchantment? Or are their objective changes to people's lives in their 40s and 50s such as poorer health, divorce and teenage children that undermine wellbeing?

Recent developments in research by economists have been influenced by the work of positive psychologists such as Martin Seligman (2002) and Ed Diener (1984, 2009). Economists have used insights into cognition to develop more sophisticated ways of modelling human behaviour, giving rise to behavioural economics (Sunstein and Thaler 2009). Instead of traditional economic models and assumptions of rational, well-informed individuals we now have rather more realistic conceptions of flawed, fallible or biased humans. To sociologists, of course, these newer more 'realistic' models of the individual are very familiar—integral to the 'sociological imagination' (Mills 1959)—as sociologists have always been interested in how everyday routines and ideologies frame interaction. For example, the paradox, popularised by Marx of how people experience 'false consciousness' whereby they are unable to make choices that further their interests. Bourdieu's work on habitus (Bourdieu 1977; Bourdieu and Passeron 1990) is similar and has been used to illustrate how able students come to experience stigma and shame about their cultural backgrounds which undermines their ability to make informed, appropriate choices about their learning careers (Cieslik and Simpson 2015). The issue of fallibility is an intriguing one and features in many of the accounts provided by my interviewees of how they struggle to make good choices about living well. It is at the heart of this notion that happiness is a riddle, something that defies easy analysis and understanding.

The influential economist Daniel Kahneman (2011) draws on experimental psychology to show how, in an effort to save time and effort, people develop cognitive short cuts and these routines hinder our ability to make informed choices about how best to be happy. When choosing one course of action over another we often have faulty memories or mistake the effects that different experiences have on our wellbeing. Hence people often struggle to make good choices when trying to live well and they fail to learn from past mistakes locking them into corrosive routines and undermining their happiness.

The synthesis of economics and psychology in recent years has led to an outpouring of texts on the nature of happiness and how best to find it. Books by Richard Layard (2005) and Paul Dolan (2015) are just two examples that use surveys and experiments to show us the many factors associated with good wellbeing. Layard's book is very popular as he succinctly summarises the many developments in positive psychology, particularly around fallibility and the challenges this poses for people seeking a happy life. Layard suggests there are seven key features associated with happiness such as family relationships, financial situation, work, community and friends, health, personal freedom and personal values (Layard 2005: 63). Hence to be happy he suggests we engineer our lives to adopt these lifestyle features and manage best we can our faulty cognitive processes. I do explore the significance of these factors for happiness with my interviewees, asking them to create maps that charted their experiences through these different aspects of their lives that were noted by Layard. However, one difficulty with such large datasets is trying to establish the relative importance of these different factors for wellbeing. Layard, for example, says that the surveys point to the significance of family relationships and income above the others. But in reality these different domains of life interact in complex ways as we age, shaping our wellbeing in different ways. And so a rationale for this current biographical project was to investigate how wellbeing emerges over time out of the intermingling of these different experiences around intimacy, children, work, health and so on. How do these different experiences interact in ordinary people's lives? Books such as Layard's that draw on survey data rarely offer us narratives or realistic depictions of how in practice wellbeing is actually experienced in 'the round'. There are further issues with large-scale wellbeing surveys around the direction of the relationships between these different factors and wellbeing. For example, does marriage or a good job lead to better wellbeing or do happier more positive people get married and secure better jobs? If surveys struggle to offer more nuanced, explanatory accounts of wellbeing then their findings do appear banal or overly normative—that to be happy one has to marry, have a satisfying well-paid job, good health and many friends. One can appreciate how some commentators such as Sara Ahmed (2010) have concluded that happiness research functions discursively or ideologically, marginalising minorities for 'failing' to conform to conventional ways of living, rendering them in the process quite miserable.

Paul Dolan (2015), like Richard Layard (2005) and Daniel Kahnneman (2011) also offers numerous examples of how we develop routines in life that distract us from attending to the many and often simple experiences that cumulatively can help us foster better wellbeing. All of these books informed by behavioural economics and positive psychology endeavours to shed light on the hidden processes behind wellbeing. These many writers claim that by offering ordinary people insights into the opaque world of happiness they can begin to change their routines and live better, happier lives. One aim of my research has been to examine such claims—whether these self-help books and small changes to one's life do actually foster new, happier ways of living? One of the riddles of happiness is that many people buy these books and are aware of the need to change their lives yet they struggle to do so. What are the barriers that people face that prevent them from adopting these life-changing techniques? Can a more critical, sociologically informed analysis of wellbeing offer us some insights into how and why this happens?

Bibliography

Ahmed, S. (2010). *The promise of happiness*. London: Duke University Press.

Aristotle. (2009). *Nicomachean ethics*. Oxford: Oxford University Press.

Bentham, J. (1776). *'Preface to the first edition', A fragment on government*. In Mack, M. P. (Ed.). (1969). *A Bentham reader*. New York: Pegasus Press, p. 45. In McMahon, D. (2006). *Happiness: A history*. New York: Grove Press, p. 212.

Bentham, J. (1989). In J. Dinwiddy. *Bentham*. New York: Oxford University Press, p. 50, cited in McMahon, D. (2006). *Happiness: A history*. New York: Grove Press, p. 219.

Bhutan. (2012). *Wellbeing and happiness: Defining a new economic paradigm*. Retrieved from http://www.2apr.gov.bt/

Blanchflower, D., & Oswald, A. (2004). Wellbeing over time in Britain and the USA. *Journal of Public Economics, 88*(July), 1359–1387.

Blanchflower, D., & Oswald, A. (2008). Is wellbeing U shaped over the life cycle? *Social Science and Medicine, 66*, 1733–1749.

Bourdieu, P. (1977). *Outline of a theory of practice*. Cambridge: Cambridge University Press.

Bourdieu, P. (2010). *Distinction*. London: Routledge.

Bourdieu, P., & Passeron, J. C. (1990). *Reproduction in education, society and culture*. London: Sage Press.

Cantril, H. (1965). *The pattern of human concerns*. New Brunswisk, NJ: Rutgers University press.

Cieslik, M., & Simpson, D. (2015). Basic skills, literacy practices and the hidden injuries of class. *Sociological Research Online, 20*(1). Retrieved from http://www:socresonline.org.uk/20/1/7.html

Deaton, A. (2012). The financial crisis and the wellbeing of America. *National Bureau of Economic Research*, 343–368. Retrieved from http://www.nber.org/chapters/c12447

Diener, E. (1984). Subjective wellbeing. *Psychological Bulletin, 95*, 542–575.

Diener, E. (2009). *The science of wellbeing: The collected works of Ed Diener*. London: Social Indicators Research Series.

Dolan, P. (2015). *Happiness by design: Finding pleasure and purpose in everyday life*. London: Penguin.

Easterbrook, G. (2003). *The progress paradox: How life gets better while people feel worse*. London: Random House.

Easterlin, R. (1974). Does economic growth improve the human lot? Some empirical evidence. In P. David & M. Reder (Eds.), *Nations and households in economic growth*. New York: Academic Press.

Fox, J. (2012). The economics of wellbeing. *Harvard Business Review*, January–February. Retrieved from https://hbr.org/2012/01/the-economics-of-well-being

Frey, B., & Stutzer, A. (2002). The economics of happiness. *World Economics, 3*(1), 1–17.

Frey, B., & Stutzer, A. (2014). Economics and the study of individual happiness. In S. David, I. Boniwell, & A. Conley Ayers (Eds.), *The Oxford handbook of happiness* (pp. 431–447). Oxford: Oxford University Press.

Graham, C., & Pettinato, S. (2002). *Happiness and hardship: Opportunity and insecurity in new market economies*. Washington, DC: Brookings Institute.

Hockey, J., & James, A. (2003). *Social identities across the life course*. London: Palgrave.

Hutcheson, F. (1725). *Inquiry into the original of our ideas of beauty and virtue*, cited in McMahon, D. (2006). *Happiness: A history*. New York: Grove Press, p. 213.

Kahneman, D. (2011). *Thinking fast and slow*. London: Penguin Books.

Kahneman, D., & Deaton, A. (2010). High income improves evaluation of life but not emotional wellbeing. *Proceedings of the National Academy of Sciences, 107*(38), 16489–16493.

Kennedy, R. F. (1968). Speech at the University of Kansas, March 18.

Lane, R. E. (2000). *The loss of happiness in market economies*. London: Yale University Press.

Layard, R. (2005). *Happiness: Lessons from a new science*. London: Penguin.
Marx, K. (1983). Alienated labour. In E. Kamenka (Ed.), *The portable Karl Marx*. New York: Penguin.
Mills, C. W. (1959). *The sociological imagination*. Oxford: Oxford University Press.
Nussbaum, M., & Sen, A. (eds) (1993). *Introduction*. The Quality of Life, Oxford: Clarendon Press.
Nussbaum, M. (2000). *Women and human development: The capabilities approach*. Cambridge: Cambridge University Press.
OECD. (2004). Is GDP a satisfactory measure of growth? *OECD Observer*, December 2004–January 2005, 246–247.
ONS. (2011). *Measuring what matters: National statisticians reflections on the national debate on measuring national wellbeing*. London: ONS.
ONS. (2015a). *Measuring national wellbeing: Life in the UK 2015*. London: ONS.
ONS. (2015b). *Measuring national wellbeing. Insights into loneliness, older people and wellbeing, 2015*. London: Office for National Statistics.
Schwartz, B. (2004). *The paradox of choice: Why more is less: How the culture of abundance robs us of satisfaction*. London: Harper Collins.
Seligman, M. (2002). *Authentic happiness: Using the new positive psychology to realise your potential for lasting fulfilment*. New York: Free Press.
Sen, A. (1999). *Development as freedom*. Oxford: Oxford University Press.
Smart, C. (2007). *Personal life*. Cambridge: Polity Press.
Smith, A. (2012). *The wealth of nations*. Hertfordshire: Wordsworth Editions Ltd.
Stevenson, B., & Wolfers, J. (2008). *Economic growth and subjective wellbeing: Re-assessing the Easterlin paradox* (NBER Working Paper No. 14282). National Bureau of Economic Research, MA, USA. Retrieved from http://www.nber.org/papers/w14282
Stiglitz, J., Sen, A., & Fitoussi, J. P. (2009). *Report by the commission on the measurement of economic performance and social progress*. Paris. Retrieved from http://www.insee.fr/fr/publications-et-services/dossiers_web/stiglitz/doc-commission/RAPPORT_anglais.pdf
Sunstein, C., & Thaler, R. (2009). *Nudge: Improving decisions about health, wealth and happiness*. London: Penguin.
Thin, N. (2012). *Social happiness: Theory into policy and practice*. Bristol: Policy Press.
UNDP. (2015). *Human development report 2015*. New York: United Nations Development Programme.
Weber, M. (1990). *The protestant ethic and the spirit of capitalism*. London: Allen Unwin.

4

Positive Psychology and Happiness

As we have seen philosophers and economists have long debated the nature of happiness and how to live well. The review of these literatures has given us insights into the complexities of wellbeing. Some philosophers suggest we should use reason to become 'craftsmen' of life or that 'good works' can be the source of virtue or that an understanding of desires can help us to live happier lives. I explore some of these issues empirically with interviewees in later chapters, examining their own solutions to this challenge of living well. For economists their concern has been with assessing the value of things and their relative influence on wellbeing. A key challenge has been how to model the complexity of these issues with mathematical techniques whilst still holding onto the messy reality of wellbeing. As we have seen with behavioural economics some have drawn on psychological theories to do this. In this chapter we discuss some of the key developments in the psychology of happiness—particularly those that can be applied to our biographical and qualitative study of wellbeing. This positive psychology has been influential in raising awareness of happiness research, helping to create a popular psychology industry around happiness, flourishing and wellbeing. One question I posed to my interviewees was whether this industry offers genuine opportunities

M. Cieslik, *The Happiness Riddle and the Quest for a Good Life*,
DOI 10.1057/978-1-137-31882-4_4

for self-development or as critics suggest (Furedi 2004; Davies 2015), merely offer superficial tips that fail to support genuine wellbeing.

Early Positive Psychology

It was during the 1960s that the beginnings of a positive psychology emerged with the work of Abraham Maslow (2013) and Carl Rogers (1990) who pondered the shift in wellbeing that was occurring in increasingly affluent Western societies following the Second World War (Boniwell 2012: 3). This work helps us understand what happiness might be by examining its sources and with it raises questions about the conditions for a good life—debates that we saw earlier in philosophical texts. Maslow's famous model of the 'hierarchy of needs' suggests that over time sources of happiness shift from the satisfaction of material needs (such as food and shelter) to more complex ones around self-esteem, respect and creativity. Some have understood this to mean that as societies or communities become more affluent then citizens increasingly come to view happiness and a good life in terms of creativity and esteem rather than material goods and subsistence (Maslow 2013). Others have used Maslow's ideas at a more personal or biographical level to suggest, for example, that the young when starting out on their lives, view financial security as key to wellbeing but that with time, for some, these concerns then give way to higher goals of self-actualisation (Veenhoven 1995)—a set of questions that I posed to interviewees for this research. Are the younger people that I spoke to more concerned with money and security than the more affluent older respondents or are their experiences more complex than these sorts of generalisations?

In particular, as Argyle (2001) documents, positive psychology and its systematic scientific approach means that it has been keen to define happiness and how it can be measured. The understanding that happiness can be broken down into three component parts (reflecting in many ways the writings of classical philosophers) is particularly useful and is one that informs this research too. Argyle suggests that happiness can be viewed as having a subjective element (subjective wellbeing) (SWB) that is made up of feelings or emotions (affect) as well as a cognitive aspect that reflects human's ability to reflect on satisfaction in life. These two subjective aspects are joined by a third, namely the objective factors such as income, employment status, health, community relations and such like that can also frame our wellbeing. Research into happiness today therefore needs to ensure that it investigates the feelings (both positive and negative as well as their intensity, duration and frequency), the internal conversations we have pondering our lives, together with the more 'structural' features of life such as family, friendships, work and leisure. The happiness mapping that I did with my interviewees, where we created large posters of the key aspects of their lives, was my attempt to capture some of these features.

From the 1970s onwards psychologists have developed increasingly sophisticated survey tools to examine the many factors that shape subjective wellbeing. There are national and multinational surveys that ask respondents to rate their wellbeing numerically (such as on 11-point scales from very satisfied to very dissatisfied) and the relative impact of different activities (such as commuting, shopping and waged work) in various domains of life on this SWB. There is even research which identifies the best ratio of good: bad experiences to ensure overall happiness in life—of about 3 good experiences to 1 bad whereas less than this is the cause of misery and indeed too many good activities can be problematic for individuals too (Fredrickson and Losada 2005). From this sort of research comes the suggestion noted by ancient philosophers that some form of balance of good and bad experiences in life is therefore important for overall wellbeing and hence suffering itself is also needed in life if one is to live a rich life of flourishing and growth (Lazarus 2003). These psychologists suggest as others have that illness, bereavement and disappointments can have with time positive consequences as they compel us to ponder the meaning and significance of things that we value. The depths of despair therefore can sometimes be a route to greater self-awareness and growth. For sociologists keen to study happiness these are important insights as they take us beyond the simplistic static models of happiness used by many contemporary sociologists. These more complex ways of experiencing happiness I explored with interviewees when they discussed their wellbeing. As we go on to see in the empirical chapters there were a surprising number of them who suggested that there were 'silver linings' to what had initially been difficult times—they had some positive experiences from disappointments in life.

As we saw with economics, psychological research into happiness has also produced some unsurprising results—for example that people commonly rank some activities (having sex, socialising and relaxing) more favourably than others such as food shopping, housework and commuting to work (Layard 2005: 15). There are some rather more surprising findings that suggest the majority of respondents in Western societies are either happy or very happy most of the time (Layard 2005: 14). And this at the time when social scientists in these societies have reported falling living standards and growing poverty and inequality (Wilkinson and Pickett 2009; Picketty 2014) which might suggest declining quality of life rather than continuing satisfaction with it.

Layard's influential text summarises research, as we saw previously, identifying the seven key factors that underpin good wellbeing such as income, family relationships, community and so on (Layard 2005: 63). He suggests that 80 % of the differences in wellbeing in different societies can be attributed to variations in just six of these factors that shape our lives (Layard 2005: 71). Layard concludes that the quality of social relationships is key to good wellbeing as they can offer a regular source of positive experiences unlike commodities or other markers of success such as homes, cars and so on. Where relationships are problematic and the source of tension and anxiety they can seriously undermine wellbeing, particularly for children and the young (Layard and Dunn 2009). For my purposes here we can use this data to examine the key factors that shape the wellbeing of interviewees—how important are these factors to my participants' happiness? Are some more important than others, does their significance vary as we age and how does class and gender interact with these sources of happiness? Although Layard documents these sources of wellbeing he also suggests that wellbeing can be enhanced significantly through the reduction of suffering brought about by problematic relationships at home, work and community. As part of this process Layard has advocated the expansion of mental health support for individuals in the UK, particularly through the use of cognitive behavioural therapy (CBT) that can offer people greater insights into wellbeing and how relationships can shape quality of life.

Happiness and Human Fallibility

A key contribution that psychologists have made to happiness studies has been around the issue of fallibility. As we saw in the chapter on economics, psychologists have been keenly interested in the errors we make due to cognitive biases in our decision-making which impact on efforts to be happy. One example of this is the difficulty we have processing the huge amounts of information across social media to make informed choices. In the past we might have had insufficient information whereas, today information overload actually hinders people's abilities to make informed choices (Boniwell 2012). For example, some studies show how our memories are often faulty—we forget much more than we ever remember and the distorted ways we often recall events hinder our efforts to live well (Gilbert 2006). We often also seek out evidence in our everyday lives that confirm our views or prejudices rather than seek out material that might question our beliefs. This process too can work against making choices that might enhance our happiness. These sorts of mental and lifestyle routines can work against efforts to change our practices into more appropriate ways of living as our surroundings change over time. Writers such as Gilbert (2006) and Deary (2014) have documented examples of how we become wedded to damaging routines that prevent us from responding to change that threatens wellbeing.

Another important strand to this work on fallibility has been the research that has spawned the growth in mindfulness and related therapies. This approach often starts off with the observation that modern Western societies have encouraged their citizens to develop a far too instrumental and rational way of living (Williams and Penman 2011). There are echoes here of the work of the Weber (1993) and Marcuse (2002) who noted the growth in instrumental rationality and the disenchantment this brings in modern societies. The argument goes that people come to live far too much in their heads—thinking through, strategising, planning, auditing and so on and as a result are distanced from other ways of thinking, feeling and living. Trying to rationalise ones way through life works for some tasks but not for many others and so we end up creating faulty ways of living based on problematic choices that undervalues emotions, feelings and spirituality. The result is that people are often very successful in moving from one goal and target to another yet still feel, despite a 'successful life' that they are stuck on a treadmill, unable to find genuine contentment. Williams and Penman (2011) draw on Eastern philosophy and psychology to offer solutions to this modern malaise of unhappiness, suggesting the need to challenge old habits and routines as well as meditation to reconnect the mind and body. I was curious to explore these issues with interviewees, and whether they too recognise this disenchantment inherent in modern life and how to overcome it. Had they too sought alternative values or faiths as a way to find a more authentic and nourishing way of living?

Martin Seligman, Authentic Happiness and Flourishing

An important strand of positive psychology in recent years has combined some of the insights of classical and Eastern philosophy with the scientific rigour of experimental research in the process demonstrating how academic work can create popular self-help guides. Figures such as Martin Seligman have tried to move beyond the distinctions between life satisfaction/hedonic happiness and more enduring notions of wellbeing captured by ideas such as Eudaimonic happiness. In his work on authentic happiness (Seligman 2002) Seligman acknowledges the role of correlates of happiness generated by previous studies but by drawing on philosophical insights into wellbeing points to the ways that individuals can actively work to cultivate a better wellbeing. By amplifying positive emotions, engaging more with everyday activities and seeking out more meaningful relationships everyone can begin to experience more positive wellbeing (Seligman 2002). With time, one can become more knowledgeable and wiser about how to live a good life of positive experiences and with it a deeper sense of one's humanity. However, Seligman does caution readers against the view that there is a simple way for quickly transforming one's wellbeing. Instead he offers a formula that denotes that

Happiness = Set point (50 %) + Circumstances (10 %) + Voluntary Factors (40 %)

He documents research that suggests that for many people a large portion of one's wellbeing has a set point so that over one's lifetime, though there may be periods of positive wellbeing (contentment) and negative wellbeing (suffering), these tend to move around a set point and after a period return to the earlier state. The origins of this set point lie in early childhood experiences and characteristics that we inherit that shape our personalities (e.g. optimists versus pessimists). Though marriage, becoming parents and success at work for example may have positive or negative consequences these circumstances only affect small changes to overall wellbeing. Some 40 % of wellbeing, however, can be changed (voluntary factors) and it is this aspect that his books are primarily concerned with. This issue of how much of a person's wellbeing can be transformed has proved to a contentious one. With various contributors suggesting that much wellbeing is predetermined by genetic factors (trait theory) where others (state theory) point to wellbeing, though being stable, can change over the life course because of life events (Veenhoven 1994). A version of these debates is that around the theory of adaptation that suggests that life events can improve or decrease wellbeing but that after a short period wellbeing then returns to its earlier position (Brickman et al. 1978). As we have seen earlier, some refer to this as the 'hedonic treadmill' thesis—that life events unbeknownst to many can only ever have a short-term impact on wellbeing and thus individuals then seek out the next goal or activity they hope will enhance their wellbeing and continues on the treadmill of life. The message behind these very popular books, and one I explore with my interviewees, is that irrespective of one's social background most people can with the aid of some psychological insights and a measure of disciplined activity change one's life for the better. However, these improvements to wellbeing may be short-lived unless individuals are vigilant about monitoring their wellbeing.

Mihaly Csikszentmihalyi and 'Flow' Experiences

One of the weaknesses of much sociological research is its preoccupation with the pathologies of modern living and neglect of positive experiences. Thus I have been forced to draw on research from other disciplines such as psychology in order to offer a more balanced analysis of wellbeing. The psychologist Csikszentmihalyi (2002) identifies what he refers to as flow experiences where one is actively engaged with a challenging task so that individuals are able to 'lose oneself' in that moment in a way which is pleasurable. The point being that such experiences are fundamental to our nature, are linked to the use of skills, sociability and learning and so are energising and life-enhancing. The more one is able to construct a life with flow experiences then the more likely one is to flourish in life and to be happy. He cites listening to music or reading, playing sport or dancing or having sex as examples of common flow experiences—though such events can vary culturally and individually so that some may encounter such events in waged work whereas many would find work quite the opposite of flow (Csikszentmihalyi and Leferve 1989). Though it appears that Csikszentmihalyi's flow events as with these other self-help texts are focused on individual development they all recognise the social aspect of flourishing so for example many of the flow cases noted by Csikszentmihalyi are collaborative social activities such as dancing or singing or performing music reminiscent of earlier anthropological and sociological studies of social rituals or social solidarity (Durkheim 2008).

Criticisms of Positive Psychology

The major contribution of positive psychology has been to focus greater attention on how we experience positive wellbeing or flourishing and how this can be promoted at an individual, community and societal level. This has been neglected by most of the social sciences that appear unbalanced as they are skewed too much towards the investigation of suffering and how we alleviate this. Though this focus on social pathologies is essential at the same time it neglects questions about the nature of what a good life or flourishing means for people and so often fails to grasp the richness of what life and being human is and could be. Though by researching and reducing suffering we undoubtedly enhance the wellbeing of individuals and communities this is not the same as researching and promoting flourishing.

Although leading figures in the field of positive psychology acknowledge the place of suffering in wellbeing and the need for a balanced approach to research (that explores the relationships between good and bad experiences in life) there are tensions and contradictions in how this area of research has evolved. Hence the effort here in this book to illustrate how a sociological approach to happiness might be insightful. My argument is that by combining some of the insights from positive psychology and its focus on flourishing with some of the traditional approaches from sociology (on suffering and the structuring of experiences) we can offer a more balanced study of wellbeing that is offered by either.

One can see the need for a balanced approach to wellbeing when interviewing people for I soon realised that individuals undertake a form of mental auditing when you ask them to talk about the quality of their lives. They weigh up the good and bad things that have occurred and so make sense of their wellbeing by comparing the difficult, problematic events with the more pleasant and satisfying. Qualitative research which produces narratives about happiness inevitably involve interviewees and researchers undertaking some form of mapping exercise that documents the spectrum of positive and negative features of someone's life. In discussions with my interviewees, it soon became apparent that people make sense of the good in their lives often through comparisons with the not so good. This need for the investigation of both positive and negative events also featured when talking to interviewees about their biographies—how they had aged and how their wellbeing had changed over time.

Interviewees also discussed the many intriguing ways that earlier experiences (in education, relationships with parents and at work) can structure their present lives, their choices and wellbeing. One needs then, as I go on to discuss in the following chapter, a way of modelling these structuring processes that stem from biographical events (using concepts such as habitus and psychoanalytical ideas) as well as those that operate to frame everyday institutional encounters and the dynamics of interactional settings (using concepts such as field, capitals and a performative approach to analysis). Such a model conjures up notions of happiness that reflect the emotional or hedonic characteristics of wellbeing but the structuring of these emotions also calls for an Eudaimonic understanding of happiness as developed by thinkers such as Aristotle (2009). This constructs happiness as a dynamic, social process whereby individuals are compelled to use practical reason to navigate their way through settings, working through their wellbeing often in competition with others. The next two chapters focus on psychoanalytical theories and sociological approaches to happiness that illustrates how we might develop these forms of analyses of wellbeing.

Bibliography

Aristotle. (2009). *Nicomachean ethics*. Oxford: Oxford University Press.

Argyle, M. (2001). *The psychology of happiness* (2nd ed.). London: Routledge.

Boniwell, I. (2012). *Positive psychology in a nutshell: The science of happiness* (3rd ed.). Milton Keynes: Open University Press.

Brickman, P., Coates, D., & Janoff-Bulman, R. (1978). Lottery winners and accident victims: Is happiness relative? *Journal of Personality and Social Psychology, 36*(8), 917–927.

Csikszentmihalyi, M. (2002). *Flow: The classic work on how to achieve happiness*. London: Harper and Row.

Csikszentmihalyi, M., & Leferve, J. (1989). Optimal experience at work and leisure. *Journal of Personality and Social Psychology, 56*(5), 815–822.

Davies, W. (2015). *The happiness industry: How the government and big business sold us wellbeing*. London: Verso.

Deary, V. (2014). *How to live*. London: Allen Lane.

Durkheim, E. (2008). *The elementary forms of religious life*. Oxford: Oxford University Press.

Fredrickson, B., & Losada, M. F. (2005). Positive affect and the complex dynamics of human flourishing. *American Psychologist, 60*(7), 678–686.

Furedi, F. (2004). *Therapy culture: Cultivating vulnerability in an uncertain age*. London: Routledge.

Gilbert, D. (2006). *Stumbling on happiness*. London: Harper Perennial.

Layard, R. (2005). *Happiness: Lessons from a new science*. London: Penguin.

Layard, R., & Dunn, J. (2009). *A good childhood: Searching for values in a competitive age*. London: Penguin.

Lazarus, R. (2003). The Lazarus manifesto for positive psychology and psychology in general. *Psychological Inquiry, 14*(2), 173–189.

Marcuse, H. (2002). *One dimensional man: Studies in the ideology of advanced industrial society*. London: Routldege.

Maslow, A. (2013). *A theory of human motivation*. London: Merchant Press.

Picketty, T. (2014). *Capital in the twenty-first century*. London: Harvard University Press.

Rogers, C. (1990). *The Carl Rogers reader*. London: Robinson Press.
Seligman, M. (2002). *Authentic happiness: Using the new positive psychology to realise your potential for lasting fulfilment*. New York: Free Press.
Veenhoven, R. (1994). Is happiness a trait? Tests of a theory that a better society does not make people any happier. *Social Indicators Research, 32*, 101–106.
Veenhoven, R. (1995). The cross national patterns of happiness: Tests of predictions implied in three theories of happiness. *Social Indicators Research, 34*, 33–68.
Weber, M. (1993). *The sociology of religion*. New York: Beacon Press.
Wilkinson, R., & Pickett, K. (2009). *The spirit level: Why equality is better for everyone*. London: Penguin.
Williams, M., & Penman, D. (2011). *Mindfulness: A practical guide to finding peace in a frantic world*. London: Piatkus.

5

Psychoanalytical Theory and Happiness

The sociologist Ian Craib was unusual in that he sought to marry the ideas of social theorists and psychoanalysts to shed light on the human condition. For him, the work of Freud and others offered a set of tools for interpreting or understanding the self rather than providing a scientific process for analysing psychological problems, their causes and cures.

> I do think that some insight into the way in which we have made something out of what was given to us at the start of our lives, and a sense of our limitations and possibilities, can make life less difficult and limit the damage that we can do to others, as well as ourselves. (Craib 2001: 5)

This hermeneutic approach he employed in his work as a therapist, drawing on psychoanalytical ideas to help others manage the suffering in their lives and aid their efforts to flourish. This use of psychoanalytical ideas enriched his sociological research and elements of this approach I adopt here in this study of hapiness.

In the social sciences however the popularity of psychoanalytical ideas have ebbed and flowed over the decades Feltham and Dryden (1992) Chancer and Andrews (2014). Much research into happiness/wellbeing tends rather to rely on empirical techniques that measure relationships

M. Cieslik, *The Happiness Riddle and the Quest for a Good Life*,
DOI 10.1057/978-1-137-31882-4_5

between factors such as income, housing status or age and numerical scores that reflect the subjective wellbeing of respondents. Traditional survey research into wellbeing therefore rests on certain assumptions about individuals, namely they are rational, decision-making machines. Whereas psychoanalytical ideas suggest that people are often conflicted about the experiences they have and how they make sense of their lives, as some part of their self, hidden from view, can influence them in ways they often struggle to understand.

As we have seen, though quantitative survey research has made great strides in showing how happiness can be correlated with health, marriage or age, at the same time, the biographical interviews I conducted also pointed to other 'hidden' processes at play. In particular many interviewees spoke of decisions they had made which had led to surprising and unexpected consequences for their wellbeing. Many spoke of how such events made them realise there were deeper aspects of their self which they often neglected but which at times did influence their wellbeing. During interviews respondents often tried to work through the links between their deeper self and everyday experiences and this was often facilitated by mapping out the biographies of individuals and charting their experiences over time. Several interviewees had undergone counselling or therapy where they had been helped to explore aspects of their lives to improve their wellbeing. Significantly, though these unanticipated experiences were unsettling and often challenging they also offered opportunities for individuals to learn about themselves and reframe views, ideas or aspirations. Although some regard psychoanalytical ideas as deterministic (seeing the behaviour and psyches of individuals as driven by early life experiences) later thinkers such as Craib use them, in relation to therapeutic approaches, to examine the creative ways that people can work through the challenges they face in life. In this way I believe that psychoanalytical concepts can offer some insights into the struggles people face living a good life that emphasises their agency as much as the ways they are structured by discourses (as suggested by Sara Ahmed 2010) or by the power of corporations and therapy culture (as suggested by Frank Furedi 2004).

I decided to incorporate some psychoanalytical ideas into the research design for this project as I had many discussions about wellbeing with respondents that seemed inexplicable. Individuals had appeared to make an appropriate choice in order to enhance their wellbeing only to find themselves experiencing a strong negative emotional reaction to this change in their lives. At other times they suggested there had been positive experiences from what they assumed would have been difficult moments in their lives.

The decision to draw on psychoanalytical ideas was also informed by the influence of these theories on popular culture. Film and TV drama as well as much fictional literature uses the unconscious, flashbacks, dreams and notions of trauma to construct popular ideas of the self and how we experience and search for a good life. Hence most people who have grown up during the latter half of the twentieth century have been shaped by a popular culture infused with Freudian ideas that furnish us with a vocabulary to explore our identities and wellbeing.

Nietzsche famously wrote in the preface to the Genealogy of Morals that we are 'strangers to ourselves' (2003). This idea that our self can be a mystery to us and that we are fallible, struggling to make sense of how to be happy are important themes that run through this book—hence its title. Psychoanalytical ideas then offer us some concepts to model how our everyday conscious deliberations and efforts to live well are influenced by our deeper self and which can be linked to earlier experiences in life. As I explore later on, many of the people I interviewed spoke of the difficulties they had reconciling themselves to these hidden aspects of their identities yet coming to terms with these features and knowing oneself more deeply was one way eventually, of living a richer, more fulfilling life. It was one way in which even the most difficult of times could eventually come to have a 'silver lining'.

Freud's Models of the Self

The three-fold model of the psyche developed by Freud allows us to conceptualise the self as a dynamic and inherently conflictual process, where people are often torn between desires and the need to conform to social norms (Elliott 2015). As individuals develop and grow the instincts that shape and constitute their 'id' such as the desires for pleasure and bonding are joined by an awareness of others and society that comprise the 'super-ego'. The tensions between fulfilling our wishes to be happy (the 'pleasure principle' from the 'id') and normative and moral expectations of others ('reality principle' from the 'super-ego') creates all manner of ambivalences and contradictions in how individuals come to make sense of their everyday lives and how they may be happy.

Freud views the development of the third part of the psyche, the 'ego' as important in allowing individuals to consciously develop ways of managing some of these tensions in the self between drives and the pressures to conform to social norms. However although the ego has executive, decision-making functions its working is complicated by the suggestion that much of the 'id' and 'super-ego' operate beneath consciousness. The conscious part of the psyche only glimpses aspects of these hidden features symbolically (such as in dreams) and they appear in proxy forms that are easily misrecognised by us when thinking through our internal worlds and past experiences. Individuals therefore, often with the help of others such as therapists have to develop creative ways to interpret their inner worlds and how they influence their wellbeing.

Freud suggested that this model of the psyche develops during the first decades of life and that the early social relationships we experience can create templates that frame how we think and relate to people and which have a profound effect on our wellbeing as we age. Hence the interest I had in exploring the early life experiences of interviewees and their influence on wellbeing. Freud discusses the development of the individual from the dependency of the suckling infant (the oral phase) to the small child becoming aware of herself in relation to her parents (the anal and then phallic stages) and then into adolescence and the awakening of more adult sexual identities (the genital stage). The 'normal' process of moving from one stage to another poses psychological challenges as the child learns to make sense of shifting relationships with others coupled to his/her physical/ mental maturation. Hence from an early age we experience disappointments, anger and envy as well as joy, pleasure and hope and these formative experiences can structure how our self works. To grow up to be a happy well-rounded individual then one needs a range of these experiences—disappointments and suffering can have an important place in our efforts to flourish (Craib 1994). Thus the argument in this book that a sociological conception of happiness must move beyond the idea that happiness is only a positive emotional experience to suggest it also involves the active balancing of good and bad experiences and emotions. Freud suggested however that at some points during these developmental stages the child may have problematic relationships with others which establishes cognitive templates (ways of thinking and feeling as well as unconscious characteristics) that can negatively influence the wellbeing of individuals as they grow to become adults. If the child is neglected, lacks love and attention or has controlling parents these sorts of experiences can shape the psyche and establish patterns of thinking and behaviour that make individuals vulnerable to poorer wellbeing. For example, some children may become anxious if caregivers neglect their needs or if they are overly controlling about children's behaviour. Children then may develop psychic templates that can endure into adulthood making them vulnerable to anxieties themselves and ways of managing these that may involve problematic choices and behaviours—such as drinking, smoking, drugs and so on. As I go onto show many of my interviewees spoke of how their childhood experiences cast a long shadow over their lives. Many had problematic relationships with parents and partners that influenced their psyches and in turn their wellbeing that suggested the need for a psychoanalytical dimension to this study of happiness.[1]

A key aspect of Freud's theory of the self is his discussion of psychic defence mechanisms which reflect the efforts, often unconsciously of individuals to manage the tensions between ones desires and social norms (Freud 1992). There are many different varieties of these such as denial, rationalisation, splitting and so on. The most notable of them is the idea of repression whereby desires or fears are blocked and actively forgotten as they transgress social norms. Freud's work into sexuality offered many famous examples of this (Freud 1901). He suggests that repression is only ever partial and so the ideas and experiences that have been internalised and forgotten can seep back into our conscious life, creating unpredictable and often disturbing emotional responses to what should be innocuous and mundane everyday events. Freud termed these episodes, slips or 'parapraxes', such as we see when we feel a sense of déjà vu but struggle to know why this is the case. One way of understanding why humour and

[1] For some people however they develop far more serious and disabling mental health problems which exceed discussions of wellbeing and whose causes often lie in complex physiological and psychological processes.

'having a laugh' can be good for wellbeing is that it allows us to explore and share in socially acceptable ways some of the tensions and contradictions that structure our selves. As we see later on, interviewees spoke of the important role of friends to a good life and laughing with friends was key to wellbeing. Freud also wrote extensively about dreams and day dreaming and how these can offer insights into the ways that our experiences have been internalised and influence our conscious life. Dreams and day dreaming then can offer symbolic representations of some of the fears and desires that we have and offer one way of interpreting some of the difficulties and also pleasures that we have in our conscious life.

'Civilization and Its Discontents'

Towards the end of his life Freud offered a pessimistic appraisal of the possibilities of happiness in modern societies. His account echoes Marx and Durkheim's depiction of 'happiness as a problem' as increasingly he felt it was experienced in superficial ways that masked more authentic ways of living. In his book Civilization and Its Discontents (2004) he suggests we have to increasingly submit to the needs of others and it is this growing cultural regulation that drives our psychic self-control and unhappiness. Freud suggests that we come to work on our own psyches in order to conform to wider cultural norms—a view that seems to prefigure the later writings of Foucault and arguments about 'technologies of the self' developed by later theorists such as Nickolas Rose (1999) and Sara Ahmed (2010). Our drives (Eros—the desire for personal growth and expressions of sexuality and Thanatos—expressions of anger, aggression and ideas around death) have to be managed to conform to the complex moral and social norms today. Hence much of modern life is about being unfulfilled, experiencing guilt about our blocked desires and seeking ways of finding enjoyment in alternative yet less fulfilling activities.

> Civilisation is built up on renunciation, how much it presupposes the non-satisfaction of powerful drives—by suppression, repression or some other means. Such 'cultural frustration' dominates the large sphere of interpersonal

> relations; as we already know, it is the cause of the hostility that all civilisations have to contend with. (Freud 2004: 44)

> The sense of guilt is the most important problem in the development of civilisation…the price we pay for cultural progress is a loss of happiness, arising from a heightened sense of guilt. (Freud 2004: 91)

I was keen to explore these dilemmas with interviewees—had they experienced this mismatch between the possibilities for happiness and pressures to conform and how did they cope with these tensions? Did coping strategies lead unintentionally to obsessive behaviours or problematic relationships that ultimately undermine wellbeing?

The Influence of Freud on Sociology

Freud's ideas have influenced other writers who have explored happiness, such as the Frankfurt School members, Herbert Marcuse and Erich Fromm. Both wished to offer a critique of consumerist societies yet proffer a more optimistic analysis than offered by Freud. These writings by the Frankfurt School are useful in illustrating the pressures that people face in capitalist societies where work is often unfulfilling and we face demands to live consumerist 'successful' lifestyles. In their books *Eros and Civilisation* (Marcuse 1987) and the *Fear of Freedom* (Fromm 2001) they explore the psychological tensions that people struggle with trying to manage the desires to live an authentic, free life (the pleasure principle in Freud's terms) whilst constrained by social norms to conform to a life of waged labour and the superficial happiness of consumerism (the reality principle). A modified more 'social' concept of repression is used by these writers to explain how individuals submit to a system that only offers a superficial, alienating existence. Hence capitalism has endured, despite its corrosive effects, as any critical consciousness that may have emerged has been neutered, not only by the power of ideologies as suggested by Marx but also by psychological processes of repression. Yet as repression is only partial, people do come to sense some of the contradictions between the desire for a freer authentic way of living and that which is offered in capitalist societies. For some, they are able to work through this fractured sense of the self in ways that draw on creativity, art, imagination and fantasy to explore a different way of existing and being happy (Marcuse 2002). For others though these tensions in their psychological and wider social life leads to more pathological adaptations that involve obsessive behaviours (Fromm 2001).[2]

The synthesising of the ideas of Marx and Freud is useful to interrogate my interviewees' experiences of wellbeing. At its simplest their testimonies suggested they were often unaware of how and why they felt good or bad about particular events. And so some notion of a deeper self or unconscious and the tensions between this and our conscious self is one way of making sense of this fallibility. In interviews we went to great lengths to establish how the experiences of happiness/sadness might be linked to past events in their lives or the complex social networks they had through work, families and leisure activities. At times we were able to establish some connections between their subjective feelings and the events that may have generated these experiences. All my respondents acknowledged that earlier events in their lives might have influenced them in one way or another—that we internalise some of these experiences and some of this emotional energy remains and influences our development and our wellbeing as we age. As I noted earlier Ian Craib explored some of these processes, acknowledging that such analyses are always interpretative and open to contestation. Others such as Bourdieu and his concept of habitus also involve such an approach positing that early experiences in life can frame psychic dispositions and the sorts of orientations to life we establish as we age (Bourdieu et al. 1999) Bourdieu famously discusses how working-class children who fail in school because of how they talk and think rather than a weakness in intellect can be left with a corrosive psychic legacy of injustice and fatalism that compounds their material disadvantages (Bourdieu and Passeron 1990). The work of Sennett and Cobb (1977), Diane Reay (2005) and my own research

[2] For some thinkers influenced by Freud such as Lacan (2006) and Zizek (1989) they suggest that the idea of a coherent self is a myth and part of the modernist project that (like happiness itself) promotes existential insecurities and in turn the sorts of consumerism that supports capitalism.

(Cieslik and Simpson 2015) offer many examples of the shame associated with growing up in poverty and how this can be internalised creating lifelong psychic barriers to social advancement.

Freud, Popular Culture and Happiness

A further justification for drawing on psychoanalytical ideas is that many of the individuals I interviewed had undergone therapeutic interventions such as counselling which they felt had been beneficial for their wellbeing. Most of these therapy sessions entailed various forms of cognitive behavioural therapy (CBT) that focus on identifying the causes of problematic thinking and behaviour and developing strategies to mitigate some of these cognitive difficulties (Mind 2015). Although influenced by psychoanalytical ideas CBT is more concerned with alleviating suffering than exploring deeper and longer term issues around meaning and sense-making that are the concerns of traditional psychotherapeutic approaches (Craib 2001: 182–186). Nevertheless, CBT, like established psychotherapy, often involves individuals talking through their current difficulties and establishing how these are linked to prior life events, enabling individuals to see how problems with wellbeing have complex biographical and psychic origins that have often been overlooked. For many interviewees the process of investigating these connections and exploring ways of changing their thinking and lifestyles proved to be beneficial and helped them to not only be happier but also to understand themselves rather more than they had in the past.

A further reason for employing some psychoanalytical ideas is the influence of these concepts in popular culture and so even if the empirical evidence for Freud ideas may be questionable, they have undoubtedly played a role in our socialisation. Hence ideas of the unconscious, repression, slips and a fractured, ambivalent self have become powerful scripts or narratives that help people to make sense of their lives and their wellbeing. All of my interviews discussed the importance of books and films to them and how they identified with particular characters. The stories people told me about their lives, identities and happiness invariably referenced these different aspects of popular culture. Admittedly, the notion of a hidden deeper self, influencing our lives is not one that just emerged with Freud, as Shakespeare, Dante and others all famously wrote of the power of emotions and the inner self to shape our lives. Yet psychoanalytical ideas have given a significant impetus to how fictional stories are portrayed in film, television, literature and art, establishing many ways for us to identify with storylines and characters and furnishing us with ways of telling our own stories. We often make sense of the difficulties and suffering in our lives through relationships with popular culture. Our expectations of a good life and the forms these may take are often framed by what we see and hear across the mass media and other social media.

Several film theorists have explored the role of psychoanalytical ideas in the cinema (Metz 1994; Mulvey 2009) and many classic films from directors, such as Bernardo Bertolucci (*Last Tango in Paris*; *The Conformist*) Woody Allen (*Manhatten*; *Deconstructing Harry*) and Alfred Hitchcock (*Spellbound*; *Psycho*; *Marnie*) appear to use Freudian ideas in their construction. In some films there is explicit use made of psychoanalytical ideas as we see in Woody Allen's many films where therapists appear. In some of Hitchcock's films such as *Spellbound* there are characters who undergo therapy, in this case the psychoanalysis being used to address the amnesia of the central male character John Ballantine played by Gregory Peck. The opening sequence of *Spellbound* introduces the audience to the role of psychoanalysis in the film,

> Our story deals with psychoanalysis, the method by which modern science treats the emotional problems of the sane. The analyst seeks only to induce the patient to talk about his hidden problems, to open locked doors of his mind. Once the complexes that have been disturbing the patient are uncovered and interpreted, the illness and confusion disappear…and the devils of unreason are driven from the human soul. (Hitchcock in Sandis 2009: 68)

In Hitchcock's films we often see characters that are troubled (neurotic or psychotic) and have some secret to be discovered, often linked to an earlier trauma. Common to many of these films is the powerful drives that make up the psyche and how individuals struggle to control these and how repression often leads to further problems for the characters. Flashbacks, dreams and the notion of the uncanny (a sense of déjà vu) are often used to render dramatically the idea that deeply repressed ideas can appear symbolically in the consciousness of characters offering the audience hints at how hidden past events have shaped the identities of protagonists. The appeal of these cinematic techniques and narratives work at multiple levels, with the audience intrigued about the hidden lives of people who appear quite ordinary yet underneath are struggling to hold themselves together (Sandis 2009).

These Freudian motifs are evident in contemporary film and TV—the Sopranos, for example, which told the story of a Mafia family in the USA has a lead character (Tony Soprano) who undergoes psychotherapy as a way of dealing with the contradictions in his life. How is it possible to lead a 'normal' family existence of loving father and caring husband whilst also being involved in the violence, murder and terror that comes with a mob life? One can also read the more recent TV series, *Madmen* (Weiner 2007) in a Freudian way to explore some of its popularity to audiences. The series tells the stories of a group of advertising executives working in New York during the 1960s and 1970s. The appeal of the central character Don Draper is that he is an intriguing figure who has a hidden past and much of the dramatic tension of the series centres on Don's efforts to conceal what he regards as a shameful history, growing up in poverty to adopted parents after his prostitute mother had died in childbirth. The narrative arc of the series shows Don struggling to hold down his job and marriage despite aspects of his past, repressed life coming back to haunt him, through his dreams and through people he knew when young. The Freudian reading here and one that hints at the popularity of the series is that it is difficult to invent a new self as our earlier experiences do imprint themselves in our psyches. The audience watches Don's life spiral out of control as his efforts to maintain his subterfuge become increasingly difficult to manage, resorting to drink, drugs and womanising as ways of coping with the contradictions in his life/self. Ultimately he comes to realise, as he works through these crises in his life, that he needs to lead a more authentic life, one that is more true to himself and his past and so he leaves his job in advertising to forge a new life for himself.

We can see how Freudian ideas about the self and happiness have come to shape popular culture. In particular, ideas that happiness might be ambivalent and contradictory are powerful ways to model how one can live. As we see later on I was keen to examine whether my interviewees used these forms of imagery to make sense of their own experiences of happiness and the good life.

Bibliography

Ahmed, S. (2010). *The promise of happiness*. London: Duke University Press.

Bourdieu, P., & Passeron, J. C. (1990). *Reproduction in education, society and culture*. London: Sage Press.

Bourdieu, P., Accardo, A., Balazs, G., Beaud, S., Bonvin, F., Bourdieu, E., Bourgois, P. et al. (1999). *The weight of the world: Social suffering in contemporary society*. Stanford, CA: Stanford University Press.

Cieslik, M., & Simpson, D. (2015). Basic skills, literacy practices and the hidden injuries of class. *Sociological Research Online, 20*(1). Retrieved from http://www:socresonline.org.uk/20/1/7.html

Craib, I. (1994). *The importance of disappointment*. London: Routledge.

Craib, I. (2001). *Psychoanalysis: A critical introduction*. Cambridge: Polity Press.

Elliott, A. (2015). *Psychoanalytic theory: An introduction* (3rd ed.). London: Palgrave.

Freud, A. (1992). *The ego and the mechanisms of defence*. London: Karnac Books.

Freud, S. (1901). *Fragment of an analysis of a case of hysteria* (Dora), cited in Thwaites, T. (2007). *Reading Freud: Psychoanalysis as cultural theory*. London: Sage, p. 146.

Freud, S. (2004). *Civilisation and its discontents*. London: Penguin.

Fromm, E. (2001). *The fear of freedom*. London: Routledge.

Lacan, J. (2006). *Ecrits: The first complete edition in English*. New York: W.W. Norton.

Marcuse, H. (1987). *Eros and civilisation: A philosophical enquiry into Freud*. London: Routledge.

Marcuse, H. (2002). *One dimensional man: Studies in the ideology of advanced industrial society*. London: Routldege.

Mind. (2015). *Making sense of CBT*. London: MIND.

Mulvey, L. (2009). *Visual and other pleasures* (2nd ed.). London: Palgrave.

Nietzsche, W. F. (2003). *The genealogy of morals*. New York: Dover Publications.

Reay, D. (2005). Beyond consciousness: The psychic landscapes of social class. *Sociology, 39*(5), 911–928.

Rose, N. (1999). *Governing the soul: The shaping of the private self* (2nd ed.). London: Free Association Books.

Sandis, C. (2009). Hitchcock's conscious use of Freud's unconscious. *Europe's Journal of Psychology, 3*, 56–81.

Sennett, R., & Cobb, J. (1977). *The hidden injuries of class*. Cambridge: Cambridge University Press.

Weiner, M. (2007). *Madmen*. New York: Lions Gate/AMC.

Zizek, S. (1989). *The sublime object of ideology*. London: Verso.

6

Sociological Approaches to Happiness

In the preceding chapters I have explored how various disciplines can help us understand the different ways that happiness is experienced today. We need to draw on a diverse range of scholarship as there are weaknesses with how sociologists have traditionally understood happiness and wellbeing (Veenhoven 2008; Bartram 2012). In this chapter I argue that mainstream sociology offers rather limited approaches to happiness research, in part driven by the way the discipline has fragmented into many different specialisms such as employment, family, education, leisure and so on. As happiness or the good life is something that cuts across many different domains in our lives, a fragmented academe with multiple literatures and research traditions hinders efforts to make sense of how happiness functions in society.

Although sociologists working in their specialist fields make important insights into aspects of wellbeing, they often fail to link this work to research in neighbouring fields leaving us with a partial or thin analysis of wellbeing. Other traditions in sociology (in a way similar to the economics of happiness) view happiness as something that is a measurable characteristic of individuals and study it through quantitative surveys. This fails to capture the relational nature of wellbeing and how it often

M. Cieslik, *The Happiness Riddle and the Quest for a Good Life*,
DOI 10.1057/978-1-137-31882-4_6

emerges collectively, rooted in different domains in life as well as through our biographies. There are also those 'sceptical approaches' that often see happiness in a narrow subjective way as 'good feeling' and regard it as a problematic symptom of a wider malaise in modern societies. These approaches neglect the many different ways that positive experiences frame our lives—the attachments we have to people, places and ideas—from the things that really matter to people that we often see depicted in great literature, poetry, art and cinema (Sayer 2011). These perspectives abstract from the complex ways that happiness can be experienced in everyday life and have hindered efforts by sociologists to study wellbeing in a more rounded way. Hence there is the need to draw on this wider scholarship to develop a more sustained empirical analysis of happiness. Yet at the same time the traditions of sociological research do offer happiness studies a useful set of tools that allows us to critically examine the experience of happiness, situating it in relation to shifting power relationships and changes in culture, policy and economics. The aim of this chapter, therefore, is to document some of the different ways sociologists have researched happiness, what we can take from these approaches and how best to combine them with insights from elsewhere to create a more robust sociological approach to the study of happiness.

When I first became interested in the nature of wellbeing around the turn of the century it was immediately apparent that few mainstream sociologists in the UK were concerned with researching happiness. There were very few articles or books on the subject; happiness was not a topic that featured in textbooks or undergraduate programmes.[1] Yet, 'being happy' struck me as a central problematic in our lives and one you would have expected sociologists to study. So I was intrigued why such a gap in research existed in sociology when so many other disciplines such as economics, philosophy and psychology had been examining wellbeing for many years. My initial literature review suggested that although projects are usually focused on questions around quality of life, the ensuing analyses produced accounts of 'ill being', pathologies and deficits (Cieslik

[1] Granted, the new Journal of Happiness Studies was launched in 2000 by Dutch sociologist Ruut Veenhoven but the majority of contributors were at the time predominantly psychologists and economists.

2015).[2] The dominant sociological mode of operation usually consists in analysing suffering and the assumption that we may understand well-being and its promotion by reducing suffering. In their preoccupation with the pathologies of modern societies many sociologists seem curiously neglectful of how we may define happiness and a good life, how these are experienced and what are the sources of these features of living. My reviews of classical and some contemporary sociological research suggested that instead of an ambitious processual approach to happiness (as one might have expected) most tended to understand happiness and well-being in narrow subjective and psychological terms—happiness as aspects of egoism, individualism or narcissism associated with the problematic, historical development of modernity. As I show, the origins of this 'happiness as a problem' approach lie in the classic writings of thinkers who established the discipline and the 'sociological imagination' which have proved at times to be an impediment to a sociological approach to happiness. To develop a 'happiness lens' as a sociologist one has to challenge the scepticism the discipline has for happiness studies and use the analytical approaches it offers as best as one can.

The Legacy of Classical Sociology

August Comte who first coined the term 'sociology' recognised that as religion declined, there were doubts that science and reason could offer a substitute for the loss of magic and meaning that came with secularism. The pursuit of happiness, he thought, might be one way of fostering solidarity in troubled times—hence the interest in researching and promoting social progress through his 'positivism', as outlined in works such as the 'Course in Positive Philosophy' (2012). Here we see the first discussion of the significance of happiness ('Bonheur' to Comte) in sociology. Ple` (2000) suggests that Comte saw positivism as a sort of 'tool kit' to help organise society and for people to live a good life. Comte wrote that positivism would only succeed in doing this if it helps to marry two different aspects of practice—reasoning/scientific endeavour

[2] Some of these arguments have been reproduced in my article (Cieslik 2015).

with underlying values, ethics and emotions. Happiness is understood by Comte then as a goal to strive for in modern societies—an ideal state of good living (for individuals and wider society) that can be achieved when individuals are helped (via the use of positivism) to understand the social and natural world around them through reasoning and informed by a set of ethics and values that also allows for a spirituality and meaning in a secular society. Of interest here is how the first sociological ideas acknowledge that for an enduring happiness we have to use reason to explore it and that our activities have to have meaning for us—they have to be connected to some deeper values or principles. As we see later on, many of the interviewees in this project were trying to achieve this more enduring sense of fulfilment with differing degrees of success. This sort of contentment in life was one of the main challenges all the interviewees faced at different points in their lives.

Happiness in the Writings of Durkheim and Marx

The writings of Durkheim and Marx offer some powerful insights into happiness in contemporary societies, yet at the same time we also see scepticism about the notion of happiness. Durkheim embarked on his research as he felt that good wellbeing was threatened by the development of modernity. In his book *The Division of Labour in Society* which was published in 1893 he charts the historical emergence of modern societies and suggests their dynamism erodes the structures of beliefs and practices (the 'conscious collective') and with it the key ways in which individuals maintain a sense of place, belonging and hence good wellbeing. In his books *Suicide* (1984) and *The Elementary Forms of Religious Life* (1912) he describes the significance of collective rituals and traditions for humans and how quality of life always resides in the quality of our social relationships. These books suggest that wellbeing is socially constructed and rooted in the moral order, rather than having psychological origins and explanations. They were also influential methodologically for later sociologists who developed larger-scale survey-based approaches to wellbeing (Abbott and Wallace 2012; Veenhoven 1999).

> The number of suicides is continually growing. We are therefore faced with a phenomenon not linked to any special local circumstance but to the general atmosphere of the social environment...This means that the decreasing happiness that the progression in the number of suicides demonstrates is the average happiness...progress does not increase our happiness very much, since this decreases, and in very alarming proportions, at the very moment when the division of labour is developing with a vitality and speed that we have never previously known. (Durkheim 2014: 193–194)

Durkheim ponders the contemporary dilemma of how one has personal freedoms and expression of individuality in dynamic societies whilst ensuring reciprocal, trusting relationships that can support good wellbeing. He is critical, therefore, of the Utilitarians such as Bentham who see happiness as an individualistic pursuit or Socialists who see wellbeing in terms of the redistribution of wealth. This is a dilemma that most of us face at some point in life—as my interviewees illustrated—how to balance one's own interests with those of others, the quest for success and material gain yet wishing also to have a family and friendships. The importance of these social relationships for Durkheim was their ability to help people sustain a sense of perspective about life and to live with moderation.

> A need or desire that know no bounds and no rules...can, for the people it affects, only be a source of permanent agony ... Is there anything more disappointing than to aim for a goal that cannot be reached because the closer we come the more it recedes? It is vain haste which does not differ from walking on the spot and which leave only sadness and discouragement in its wake. (Durkheim 1973: 93, quoted in Vowinckel 2000: 454)

Durkheim touches on the modern problematic of how to manage expectations in dynamic societies that are able to constantly expand the range of goods and services available to people—creating what modern psychologists call the 'hedonic treadmill'—an elusive lifestyle and happiness that is always just out of reach. His prescription for better wellbeing and a good society lies not necessarily in religious beliefs as in the past but in the maintenance of complex networks amongst individuals. Not only does the expansion of the division of labour provide the employment and material resources for a good life but also offers moral regulation—the limiting of desires that could be corrosive for personal happiness and wider social cohesion. As we go onto see in the later empirical chapters, Durkheim's ideas are still relevant today—for the density and quality of people's relationships were often linked to their wellbeing. For those who were unemployed or lonely, they tended to have poorer wellbeing that illustrates how we are inherently social creatures and depend on others for a good life.

Alienation and Happiness

In contrast to Durkheim, Marx and Engels were more concerned with how economic relationships shaped wellbeing and a 'good society'. Drawing on the work of Aristotle, Rousseau, Hegel and others, they suggest good wellbeing resides in the freedom to use one's labour and skills in creative ways to enable flourishing (see Giddens 1971: 6–17). In their earlier work, such as the Paris Manuscripts and the German Ideology, they discuss how control over one's labour and a balanced life is needed for an authentic wellbeing.

> As soon as the distribution of labour comes into being, each man has a particular, exclusive sphere of activity, which is forced upon him, and from which he cannot escape. He is a hunter, a fisherman, a shepherd, or a critical critic, and must remain so if he does not want to lose his means of livelihood; whereas in communist society, where nobody has one sphere of activity but each can become accomplished in any branch he wishes, society regulates the general production and thus makes it possible for me to do one thing today and another tomorrow, to hunt in the morning, fish in the afternoon, rear cattle in the evening, criticize after dinner, just as I have a mind, without ever becoming a hunter, fisherman, shepherd, or critic. (Marx and Engels 1971: 63)

When we see the pressures that many of my interviewees were under at work with increasing specialisation of tasks, little autonomy and long hours undermining a balanced, happy life—we can see the continuing relevance of Marx and Engels' writings. Sadly for happiness researchers, though they occasionally ponder the possibility of happiness in modern societies, Marx and Engels like many contemporary sociologists soon conclude that good wellbeing is almost impossible under capitalism. The way that work is organised and how economic relations shape wider society (such as families, the legal system, property, leisure and wider culture) creates myriad forms of powerlessness that are inimical to any meaningful flourishing. Hence we have to use Marx 'creatively' to study positive or nourishing experiences in modern societies. For Marx's analysis of the pathologies of modern societies helped establish an enduring template for sociological analysis rooted in the examination and alleviation of suffering rather than an investigation of wellbeing per se.

The concept of alienation is at the heart of the Marxist theory of capitalist societies and illustrates how Marx and Engels viewed the prospects for good wellbeing. In a famous passage in the Paris Manuscripts, Marx writes that a person

> does not affirm himself but denies himself, does not feel content but unhappy, does not develop freely his physical and mental energy but mortifies his body and ruins his mind. (Marx 1983: 139)

The concept of alienation illustrates the different dimensions of poor wellbeing, (alienation from other workers, the product of ones labour, the natural world and from ones creative essence as a human). However, if we turn this formulation on its head we can explore how flourishing might be achieved so that people: need to have good relationships with other people; have the opportunity to create things of use; have good relationships with nature; and opportunities to learn and be creative. Interestingly, such a typology has influenced psychologists such as Csikszentmihalyi (2002) and economists such as Amartya Sen (1999) and their thinking on flourishing in modern societies. Marx and these later thinkers identify some of the elemental features of being human such as creativity and play and how we need these to grow and live well. Indeed, the people I interviewed who tended to be happier than others were those who had learnt, despite the pressures of modern living, to set aside time for these nourishing activities.

Happiness and Ideologies

A key debate in happiness studies centres on the ideological nature of popular notions of happiness, as fun, pleasure or escapism that divert attention away from the underlying processes that frame wellbeing. Many sociologists focus their research on how these simpler notions of happiness (e.g. as hedonism) are used by corporations to promote consumer capitalism. We see the origins of this approach in Marxist critiques of Bentham's utilitarianism and religious doctrines. Marx suggested that both approaches proposed mistaken routes to a shallow or artificial happiness:

> The abolition of religion as the illusory happiness of the people is the demand for their real happiness. The demand to abandon illusions about their real condition is a demand to abandon a condition which requires illusions. The criticism of religion is thus the germ of the criticism of the vale of tears of which religion is the halo. (Marx quoted in Giddens 1971: 7)

For Marx and Engels modern societies create superficial, individualised ways of living and of being happy that distract us from more meaningful ways of living and of relating to one another. This approach to wellbeing of 'happiness as a problem' has proved influential for subsequent sociologists and helps to account for the relative neglect of happiness studies in mainstream sociology.

I argue that it is time to move on from this overly pessimistic approach that closes down further analysis of happiness and instead develop a more ambitious and balanced engagement with how wellbeing is experienced today. Although people are constrained by their experiences at work and the need for regular incomes to support their consumer lifestyles, at the same time much of what they do involves efforts to live well that are often under-researched. Such an argument, however, has ranged against it many classical and influential thinkers in sociology—not least the work of Max Weber.

Max Weber on Happiness

For our purposes here, we can see that Weber's key contribution to happiness studies is in his interest in the role of meaning and values in life. In his studies of different religions (1993) he notes that our mortality (the fear of death and the inevitability of suffering) led to great efforts to make sense of life and to give it meaning. Through the adoption of religious values individuals acquire sets of principles about how to live well, how to make difficult choices (have an ethical and moral framework) and are helped to have good wellbeing. Weber's study into different religions illustrated how people are offered various approaches to living well, to manage suffering and insecurity and that these have evolved over the centuries. Weber's famous account of the rise of capitalism (1990) relies on the existential insecurities of Protestant sects and desire for salvation as a driver of the ingenuity and hard work that fired up the emergence of Western capitalism. Such sects then came to see an ascetic way of life and hard work (and shunning of simple pleasures and happiness), and the material success this brings as signifiers of salvation. With time the idea that hard work will be rewarded with happiness in the afterlife is replaced with happiness in this life. We are left with this powerful set of ideas and values that underpin Western societies around 'The Protestant Ethic' that views happiness and a good life as one intrinsically associated with sacrifice and deferred gratification and measured by worldly material goods. This adherence to these values of hard work and sacrifice were ones that all of my interviewees had been socialised into when growing up and so most associated a successful, happy life with affluence and consumerism. Yet as we see later on many did become disenchanted with this way of living and the values that promoted it and so my interviewee's stories about happiness involved their efforts to lead different alternative lives. Although such shifts in how to live were not easy to make and were often linked to some critical moment or a crisis of wellbeing that the individuals experienced.

Weber echoes some of Comte's concerns about the decline in religious values in modern societies and the need for ethical principles that would support wellbeing. Like Marx's work on alienation, Weber too was aware of the need for people to develop their talents if they were to flourish.

Weber wrote engagingly of people's 'inner charisma' and the ability to learn, grow and create. These ideas around a 'calling' or being compelled to seek out ways to flourish have been developed by later thinkers (Taylor 1991; Csikszentmihalyi 2002) and were ones I discussed with interviewees—about the need to explore and develop their potentials. Yet at the same time, Weber, like Marx, felt that modern societies often offered few opportunities for people to develop their potential. In particular, for Weber it was the growth in scientific thinking or rationalisation, in employment and wider social relations that undermined wellbeing. For Gerth and Mills, Weber was pessimistic about the prospects of happiness as the demands of modern life meant that people became overly specialised as they tried to adapt to the rationalised modern ways of living—the so-called 'specialists without spirit',

> That freedom as carried by charisma is doomed is evident by his nostalgic remarks concerning the French Revolution…For Weber the potentially charismatic quality of man stands in tension with external demands of institutional life…capitalism is the embodiment of rational impersonality; the quest for freedom is identified with irrational sentiment and privacy. Freedom is at best a tarrying for loving companionship and for the cathartic experience of art as a this-worldly escape from institutional routines… In this conception of freedom as a historically developed phenomena, now on the defensive against capitalism and bureaucracy, Weber represents humanist and cultural liberalism rather than economic liberalism … Weber's concern (is) with the decline of the cultivated man as a well-rounded personality in favour of technical expert, who, from the human point of view is crippled. (Gerth and Mills 1982: 72–73)

Weber feared the decline in religion and the rise of science and rationalisation would lead us all to become overly specialised in all manner of aspects of life and hasten, 'a disenchantment with life' a key impediment to a meaningful, balanced and happy way of living. Weber's insights have much resonance for happiness researchers today. His worries over instrumentalism seem very modern—all of us at some point have made 'to do' lists and become fixated on targets to only realise much later that such ways of working are often inefficient and undermine our wellbeing. Our lives become dominated with targets and lists and we lose sight of the processes or the journeys that we take that are usually more important (as they are a source of wellbeing) than the outcomes. I was curious, therefore, to see the extent of this disenchantment amongst interviewees and whether some were able to cope with this very modern malaise better than others.

Simmel and Happiness

Georg Simmel offers happiness researchers some insights into researching wellbeing, even though, like other founding figures, he often documented the pathologies of modern life. His famous paper on the 'Metropolis and Mental Life' (1950) documents the threats that increasingly complex social networks and interactions pose to the quality of life of citizens. He spoke of the 'tragedy of culture'—how people might pursue frivolous ways of living and of being happy as they are over-powered by the frenetic demands of urban living. Yet at the same time he suggests city life allows individuals opportunities to develop more complex subjectivities and wellbeing because of the myriad of new relationships that it offers to city dwellers. What happens with time is that people become more cognitively sophisticated and are able to construct complex identities and social relationships that offer new possibilities for living well. These are interesting ideas about happiness today and raise some key questions about the role of complex networks in framing wellbeing. Methodologically, Simmel is also interesting for happiness researchers advocating approaches quite different from classical theorists and much happiness research today. He laments the over-specialisation of the social sciences, arguing instead for a more small-scale, holistic approach to research that focuses on psycho-social characteristics of people that involve drives, desires and needs such as love, respect, trust and compassion (Zingerle 2000). He recognised one of the key challenges that faces happiness researchers today—that wellbeing is something that people do through many life domains yet the fragmentation of the social sciences works against such holistic research.

We can take much from these classical writings in our attempt to develop a sociology of happiness. The radicalism of these authors in searching out the effects of unequal power relations is key to a 'critical happiness studies' that views the nature and experience of happiness as a contested one. Hence we can employ a view of society where different groups and individuals struggle over what happiness and a good life can be. This can mean lovers trying to agree on how best to live, or parents' doubts over raising their children to corporations using happiness research to sell the latest products. Their methodological injunctions to link people and processes, macro-micro and structure and agency should also be at the heart of the sociology of happiness. And we should also be mindful too of their warnings of the fallibility of individuals, whose daily routines often blind them to the processes that come to shape their life chances. The sociological imagination is a powerful one in showing how in everyday life we often focus on the immediate and the tangible—seeing value in things—rather than the complex social relations that give rise to such value. Yet the way that these classic works frame the sociological imagination as one focused on the social construction of the self amidst inequality and division also shapes a sociology that has blind spots over the nature of happiness. Despite the insights of Weber and Simmel, these nineteenth century traditions of documenting the suffering in modernity rather than the desires for and possibilities of growth and development have narrowed the vision of sociological research. And the emphasis on the structural origins of social experience and how individualism distorts our faculties (as we see in notions such as false consciousness, the division of labour and social facts) distracts us from studying positive aspects of life, the creative powers of individuals and groups and their struggle for a good life.

Modern Sociology and Scepticism About Happiness Research

If we trace the development of sociological thinking from the nineteenth century to the modern day there continues to be scepticism about happiness research. In this section I briefly illustrate some modern versions of this scepticism. These accounts echo many of the ideas of foundational texts, for example that superficial notions of happiness (as good feeling) are used as marketing tools to generate profits for corporations and also function ideologically to help 'de-politicize' individuals. Again the study of happiness becomes an analysis of suffering as it is argued there is little space in capitalist societies for people to explore more authentic, enduring and political conceptions of wellbeing—such is the power of capitalism to shape cultures and lifestyles. These modern analyses extend earlier research by showing that these problematic notions of happiness have become much more prevalent through popular cultures and more deeply embedded in the psychology of individuals. They have become much more pernicious, disseminated through new media and legitimised via 'scientific research' and popular psychology.

The aim of this book has been to explore the experiences of happiness in what I believe is a rather more balanced way than is often the case with sociology. Some of these sceptical writings are useful in helping to understand how happiness may function in problematic ways in the lives of interviewees. For example the pressure to 'appear happy' at work or in our relations with others can be a source of oppression that undermines wellbeing. At the same as I have argued elsewhere (Cieslik 2015), these accounts are rather one sided and as my empirical data suggests, happiness can be experienced in far more complex and positive ways by ordinary people.

The Frankfurt School and Happiness

The work of the Frankfurt School theorists, Marcuse (2002) and Adorno (1941) typifies the way that many modern sociologists understand the functions of happiness and it has proved very influential as we see with more recent iterations (Morgan 2014; McKenzie 2015; Davies 2015; Frawley 2015). For Marcuse, the rise of the mass media and expansion of consumption were linked to simplistic notions of happiness—joy, fun and pleasure—part of a dominant culture and values that promoted conformity and uncritical ways of thinking—'one dimensional thinking'. Happiness is used to impose false needs on individuals that blind people to richer more authentic ways of living.

> We may distinguish both true and false needs. 'False' are those which are superimposed upon the individual by particular social interests in his repression: the needs which perpetuate toil, aggressiveness, misery and injustice. Their satisfaction might be most gratifying to the individual, but this happiness is not a condition which has to be maintained and protected if it serves to arrest the development of the ability (his own and others) to recognise the disease of the whole and grasp the chances of curing the disease. The result then is euphoria in unhappiness. Most of the prevailing needs to relax, to have fun, to behave and consume in accordance with the advertisements to love and hate what others love and hate, belong to this category of false needs. (Marcuse 2002:7)

One aim of this project was to explore the lives of ordinary people and the ways that consumption influences their wellbeing. So one set of questions I explored with my interviewees as a test of some of these claims by Marcuse and others, focused on their purchase of goods and services and their effects on wellbeing and whether such consumerism undermines the ability for political consciousness. For Adorno popular culture (such as TV, cinema and music) acted as a sort of safety valve—allowing people to express their frustrations about their lives (as we are bombarded with images of elusive happiness) whilst closing down ways for people to challenge this disappointment and oppression. If people are happy with consumer lifestyles and the sentimental emotions this generates then they do not rise up and protest.

> When an audience at a sentimental film or sentimental music becomes aware of the overwhelming possibility of happiness, they dare to confess to themselves that the whole order of life ordinarily forbids them to admit, namely, that they actually have no part in happiness…the scant liberation occurs with the realisation that at least one need not deny oneself the happiness of knowing that one is unhappy and that one could be happy…They consume music in order to be allowed to weep. They are taken in by the musical expression of frustration rather than by that of happiness…It is catharsis for the masses but catharsis which keeps them all the more firmly in line. One who weeps does not resist any more than one who marches. Music that permits its listeners the confession of their unhappiness reconciles them, by means of this "release", to their social dependence. (Adorno 1990: 313)

Marcuse (2002) and Adorno (1941) describe a post-war American society whose affluence and apparent freedoms conceal a deep malaise. It is one where citizens are distracted by superficial ways of living and are unable to imagine more enduring and authentic ways to live. Society is organised using technology, science, mass production and market competition and seems the embodiment of reason, serving the interests of all. Efforts to critique this system and to visualise alternatives become increasingly difficult as people come to identify through their daily practices and their psychological dispositions with the values, morals and ethics of modern capitalism—and a fixation on 'being happy' which involves simplistic, good, subjective feelings. The question I pose to my interviewees is whether this is an accurate depiction of their lives? As people age, do some become aware of the shallowness of rewards from consumerism or have they become duped as Marcuse (2002) and Adorno (1941) argue? Is it possible to balance some involvement in these consumerist lifestyles whilst also pursuing a quest for a deeper authentic wellbeing that is also a critical and politically aware way of living? Do some people as Aristotle and Eastern philosophers suggest work on their wellbeing and become more sophisticated in how they live and flourish?

The Problem of Happiness: Discourses, Loss of Community, Narcissism and Emotional Labour

The view of 'happiness as a problem' that we saw with Frankfurt School thinkers is echoed in many other later studies in both the USA and UK and helps cement the scepticism that sociologists have about happiness studies. Here I identify a few examples that raise some interesting empirical questions that I go on to explore with my interviewees. For example, Sara Ahmed (2010) suggests that popular constructions of happiness circulate as discourses and are thus implicated in how power functions to set boundaries, to marginalise and moralise. Happiness or 'being happy' in modern societies has become associated with normative ways of living (e.g. as straight, white, middle class and male) and so innocuous representations of happiness function to marginalise minorities. Other sociological research that deploy notions of happiness are those interested in a 'loss of community' (Young and Wilmott 1962; Bellah et al. 1996). These have focused on how people have become more insular or privatised as employment and class cultures change. There is a decline in civic engagement and community activities as people are happy in more inward-looking consumerist lifestyles that favour the home over the neighbourhood. Christopher Lasch (1979) in the USA and Frank Furedi (2004) in the UK continue this focus on the increasingly privatised and inward-looking character of life today in the discussion of contemporary narcissism.

> One of the most distinct features of our emotional script is its celebration of happiness and contentment…The emphasis which our emotional script attaches to feeling good about oneself is a distinct feature of contemporary culture. It is underpinned by an outlook that regards the individual self as the central focus of social, moral and cultural preoccupation…Since feeling good is regarded as a state of virtue, forms of behaviour that distract the individual from attending to the needs of the self, are frequently devalued. Consequently, traditionally held virtues such as hard work, sacrifice, altruism and commitment are frequently represented as antithetical to the quest of the individual for the feeling of happiness. (Furedi 2004: 31)

Furedi focuses attention on the rise of popular psychology and a language of emotions framing our understandings of everyday life. The emergence of a 'therapy culture' promotes a narcissistic, inward-looking sensibility where people have become obsessed with being happy which paradoxically is counterproductive as it produces vulnerabilities and anxieties about one's wellbeing. This way of living is corrosive, Furedi (2004) tells us, as it hinders our engagement with the world, makes us dependent on therapy professionals and generates superficial, instrumental conceptions of wellbeing and flourishing.

This construction of the 'problem of happiness' has also been developed by some working in the sociology of emotions. Hochschild (2003), for example, depicts happiness as a performative process whereby people work on their bodily self (surface acting) and emotional selves (deep acting) to meet the demands of others. Happiness is seen as positive emotional displays rather than the more complex ways that philosophers have seen happiness as the ethical, evaluative conduct that emerges from the effort to balance the good and bad in one's life. Like other sociologists before her Hochschild tends to see happiness as problematic as these performances of happiness illustrate how modern societies have come to colonise ever more of the self. Alienation has become more insidious today requiring not just the selling of ones labour but also complex emotional techniques—the learning of, and compliance with, so-called feeling rules if we are to succeed in employment and our personal lives.

These depictions of happiness do paint a bleak picture of modern living and so I am keen to examine whether they help us to understand the happiness experienced by my interviewees. Have they become increasingly insular and privatised as they value fun and pleasures rooted in the domestic sphere rather than more traditional social pursuits outside of it? Have my respondents become fixated on 'being happy' which does more to cause anxiety than improve wellbeing? Do they have to sustain complex performances of happiness in their personal lives and at work that threatens their wellbeing. Or can these forms of acting empower some individuals that have positive consequences for their wellbeing?

Contemporary Sociology, Wellbeing and Happiness

Despite a general reticence amongst many sociologists to engage imaginatively with the issue of happiness there have been some recent studies that explore wellbeing in sophisticated ways (Frawley 2015; Hyman 2014, Jugureanu et al. 2014; Abbott and Wallace 2012). Sociologists in the areas of work, leisure (Rasmussen and Laumann 2014) health and development for example have all been keen to explore how people's particular experiences in institutions such as employment (unemployment and underemployment) impact on wellbeing or how events such as illness can undermine good wellbeing.[3] The Dutch Sociologist Ruut Veenhoven has

[3] This review is illustrative rather than exhaustive—there are many other studies I could have discussed to show how sociologists during the latter half of the twentieth century often saw wellbeing

been at the forefront of this work on subjective wellbeing exploring how we can conceptualise and research happiness sociologically (1984, 1999, 2004, 2008). He has undertaken many quantitative studies that have examined issues popularised by economists and psychologists around the sources or correlates of wellbeing such as income, aging, culture and so on. There are many other sociologists working in specialist fields who also conduct surveys into wellbeing or quality of life. For example the British Household Panel Survey offers insights into different activities and their contribution to wellbeing (Gershuny and Yee Kan 2008) and there have been various British and European surveys into health and wellbeing and also the wellbeing of young people (Children's Society 2015; Pollock 2015; Goswami and Pollock 2015).

These studies do offer some interesting insights and pose questions that I explore with my interviewees, such as the impact that changing health (Nettleton 2013), employment (Burchell et al. 1999), welfare policies (Walker and John, 2012)and work-life balance (Roberts, 2007) have on people's wellbeing (Larkin 2013; Lunau et al. 2014; Warren 2004). Sociologists have tended, however, to follow the lead of other disciplines and research wellbeing using large scale surveys that abstract from the everyday experiences of happiness and the interpersonal processes that underpin it. Or following the more traditional approach to wellbeing the studies concentrate on the pathologies of modernity (Edgell 2012; Brint 2015; Standing 2014).

Researching Happiness: A Cross-Disciplinary, Biographical Approach

When I first began this project in the early years of this century there were no sociological texts that offered guidance on how to research happiness using biographical or qualitative methods. As I write in 2016, not much has changed and so I have had to draw inspiration from other disciplines and areas of research in sociology to help me design this project.

in narrow terms through a focus on the problems of pleasure—see for example (Davies 2015; Frawley 2016).

I hope what follows will be of some use to you if you are designing your own investigation into happiness or wellbeing.

How one defines happiness will shape the nature of any research project and so my use of a multidimensional definition called for a complex research design. Drawing on classical philosophy and notions of Eudaimonia/Hedonia (Aristotle 2009) I understand happiness broadly to include fleeting emotional experiences (the hedonic aspects of pleasures and suffering) and also happiness as a longer-term social and biographical process (the Eudaimonic aspect). Such assumptions implied a methodology that could explore different positive/negative emotions and meanings and how these changed over the life course—hence the use of in-depth qualitative, biographical interviews.[4] As part of this process I created with interviewees a biographical timeline or life grid where we charted changes (or transitions) to biographies in key domains of their life such as employment, families, relationships, education, leisure and so on. As in previous research I have conducted (Cieslik and Simpson 2015) I was keen to explore key moments of change (such as marriage, ill health, becoming a parent, divorce and so on) and their significance for wellbeing. As philosophers have noted the inherently social nature of happiness—one that is co-produced through our many social relationships I also undertook a mapping exercise with each interviewee that charted their social networks through domains (such as friends, family and work) and how these contribute (negatively or positively) to wellbeing. As economists and psychologists have noted the varying contributions of these domains to wellbeing (Layard 2005: 63) I too was able to explore these relationships at a micro level using interpretative techniques. The data generated was partly analysed using computer software (Nvivo) and also via traditional cut and paste thematic analysis. I also tried to capture some of the shifting nature of happiness by asking respondents to complete a diary where they documented their positive and negative experiences over a period of several days. The respondents were asked to comment on the duration, intensity and frequency of these events and their significance for their

[4] I interviewed 19 people over several years and many were interviewed multiple times. The research conformed to the usual ethical procedures and principles of informed consent as outlined by the British Sociological Association. As part of this approach all interviewees have been anonymised and some aspects of their identities have been disguised to protect their identities.

wellbeing—techniques that have been employed by some psychologists (Argyle 2001).

The research was informed by aspects of critical realism (Sayer 1992; Archer 1988) and grounded theory (Corbin and Glazer 1998) and so I used pilot interviews during 2009–2010 to explore the usefulness of these definitions of happiness and the mapping techniques for data generation. As most happiness/wellbeing research employs quantitative methods I also asked interviewees to complete a series of standard questionnaires (Hoggard 2005: 23–28) that allowed me to generate some numerical data on different aspects of wellbeing. These questions are similar to ones used by the ONS (201 5a) in the National Measuring Wellbeing Programme and probed life satisfaction, meaning in life and various emotional experiences. I was able to contrast the questionnaire responses to the more nuanced narratives about wellbeing and which allowed me to reflect on some of the weaknesses/advantages of each approach.

My background as a sociologist meant that many of the research questions would reflect the key concerns that sociologists have with the way that power operates to shape identities, life chances and hence wellbeing. I used concepts from Pierre Bourdieu's work (Bourdieu and Passeron 1990; Bourdieu 1977) such as habitus, practice, field and capitals to theorise how early life events frame our ideas and resources. I was keen to explore how class and gender relationships influence identities and (dis)advantages as people move through life and the effect these have on wellbeing. In recent years the restructuring of work and welfare reforms have had profound effects on living standards and this was one area I investigated with participants. To interrogate the way that people adapt or manage this structuring of their lives I also employed other concepts from Erving Goffman (1990) and his dramaturgical approach that allows one to examine the creativity of individuals and the way that happiness functions in a performative way. I also drew on Arlie Hochschild's work (2003) to examine with interviewees this feature of happiness as something that one works on emotionally at various levels of the psyche.

One of the difficulties of developing a sociological approach to happiness, as I discuss above, is that sociologists, by focusing on the problems of modernity, seem to have an inherent bias against happiness studies. There are few thinkers in the discipline that explicitly set out to research the positive experiences that people have in life (though see Thin 2012). Thus when trying to document the good aspects of wellbeing and their sources I have had to 'turn sociological concepts on their heads'. Hence I have used Marx's concept of alienation in its form as non-alienation to assess how individuals have good wellbeing (1983). In this formulation Marx would see flourishing in terms of; good relationships with the natural world; the ability to foster good friendship networks; the opportunity to learn and develop skills; and the opportunity to have some control at work and chance to produce things of value to oneself (Marx 1983). Sen and Nussbaum's writings on capabilities (1993) and also Csikszentmihalyi's research into 'flow' (2002) have similarities to Marx's thinking here and I use these to assess how different aspects of life can have positive or negative consequences for wellbeing.

A final key theme that emerged from the literature and informed the research was around the issue of fallibility. In their differing ways, psychologists, sociologists and psychoanalytical theorists all acknowledge how people have faulty memories and insights and often struggle to make informed choices in life. People develop routines and cognitive short-cuts that hinder their ability to appraise situations and make choices to live well (Kahneman 2011). For some, there are aspects to our selves (desires and fears) that are hidden from us yet can influence our wellbeing (Freud 1992). In interviews I tried to be sensitive to these issues, where people discussed their routines, symbolism of events and where there were unexpected or surprising emotional responses to situations. In initial and follow up interviews (most participants were interviewed twice, 3–6 months apart) I took time to establish links between experiences that were significant for wellbeing and earlier life events. This was aided by conducting interviews in the homes of respondents and by asking the participants to draw on images and artefacts that had personal significance for them. By doing this I was able to explore how events or experiences had become internalised and were psychologically significant for interviewees with a lasting role in their wellbeing. In this way I was able at times to unpack some of the different layers of meaning that experiences had for people's wellbeing. Some commentators suggest that fallibility has a key role in shaping poor wellbeing as routines we slowly develop through life blind us to techniques that could help us live better.

Thus some questioning focused on their awareness of these processes and whether they had drawn on therapeutic support or self-help techniques such as Mindfulness (Williams and Penman 2011) to enhance their wellbeing.

Bibliography

Abbott, P., & Wallace, C. (2012). Social quality: A way to measure the quality of society. *Social Indicators Research, 108*(1), 153–167.

Adorno, T. (1941). On popular music. *Studies in Philosophy and Social Science, New York: Institute of Social Research, IX*, 17–48.

Adorno, T. (1990). On popular music. In S. Frith & A. Goodwin (Eds.), *On record* (pp. 301–314). London: Routledge.

Ahmed, S. (2010). *The promise of happiness*. London: Duke University Press.

Archer, M. (1988). *Culture and agency: The place of culture in social theory*. Cambridge: Cambridge University Press.

Aristotle. (2009). *Nicomachean ethics*. Oxford: Oxford University Press.

Argyle, M. (2001). *The psychology of happiness* (2nd ed.). London: Routledge.

Bartram, D. (2012). Elements of a sociological contribution to happiness studies. *Sociology Compass, 6*(8), 644–656.

Bellah, R., Madsden, R., Sullivan, W., Swidler, A., & Tipton, S. (1996). *Habits of the heart: Individualism and commitment in American life* (2nd ed.). London: University of California Press.

Bourdieu, P. (1977). *Outline of a Theory of Practice*. Cambridge: Cambridge University Press.

Bourdieu, P., & Passeron, J. C. (1990). *Reproduction in education, society and culture*. London: Sage Press.

Brint, K. (2015). *Sociology, work and industry* (5th ed.). Cambridge: Polity.

Burchell, B., Day, D., Hudson, M., Ladipo, D., Mankelow, R., Nolan, J., et al. (1999). *Job insecurity and work intensification: Flexibility and changing boundaries*. York: Joseph Rowntree Foundation.

Children's Society (2015). *The Good Childhood Report 2015*. London: Children's Society.

Cieslik, M., & Simpson, D. (2015). Basic skills, literacy practices and the hidden injuries of class. *Sociological Research Online, 20*(1). Retrieved from http://www:socresonline.org.uk/20/1/7.html

Cieslik, M. (2015). Not Smiling but Frowning: Sociology and the Problem of Happiness. *Sociology, 49*(3), 422–37.
Comte, A. (2012). *Cours de philosophie positive, leçons 46–51*. Paris: Hermann.
Csikszentmihalyi, M. (2002). *Flow: The classic work on how to achieve happiness*. London: Harper and Row.
Davies, W. (2015). *The happiness industry: How the government and big business sold us wellbeing*. London: Verso.
Durkheim, E. (2014). *The Division of Labour in Society*. New York: Free Press.
Edgell, S. (2012). *The Sociology of Work*. London: Sage.
Frawley, A. (2015). *Semiotics of Happiness: Rhetorical Beginnings of a Public Problem*. London: Bloomsbury.
Frawley, A. (2016). *Semiotics of happiness: Rhetorical beginnings of a public problem*. London: Bloomsbury.
Freud, A. (1992). *The ego and the mechanisms of defence*. London: Karnac Books.
Furedi, F. (2004). *Therapy culture: Cultivating vulnerability in an uncertain age*. London: Routledge.
Gershuny, J., & Yee Kan, M. (2008). Gender and time use over the life course. In M. Brynin & J. Ermisch (Eds.), *Changing relationships: Advances in sociology* (pp. 146–160). London: Routledge.
Gerth, H. H., & Mills, C. W. (1982). *From Max Weber: Essays in sociology*. London: Routledge, Kegan Paul.
Giddens, A. (1971). *Capitalism and modern social theory*. Cambridge: Cambridge University Press.
Goffman, E. (1990). *The Presentation of Self in Everyday Life*. London: Penguin.
Goswami, H., & Pollock, G. (2015). *Correlates of mental health and psychological well-being of the European youth: Evidence from the European quality of life survey* (Working Paper). Dept. of Social Sciences, Manchester Metropolitan University.
Glazer B. G., & Corbin J. (1998). *Basics of Qualitative Research: Grounded Theory Procedures and Techniques* 2nd ed. London: Sage.
Hochschild, A. (2003). *The managed heart: The commercialisation of human feeling* (2nd ed.). Berkeley: University of California Press.
Hoggard, L. (2005). *How to be happy*. London: BBC Books.
Hyman, L. (2014). *Happiness: Understandings, narratives and discourses*. London: Palgrave.
Jugureanu, A., Hughes, J., & Hughes, K. (2014, May 2). Towards a developmental understanding of happiness. *Sociological Research Online, 19* .Retrieved from http://www.socresonline.org.uk/19/2/2.html10.5153/sro.3240
Kahneman, D. (2011). *Thinking fast and slow*. London: Penguin Books.
Larkin, M. (2013). *Health and wellbeing across the life course*. London: Sage.

Lasch, C. (1979). *The culture of narcissism: American life in an age of diminishing expectations*. New York: Warner Books.
Layard, R. (2005). *Happiness: Lessons from a new science*. London: Penguin.
Lunau, T., Bambra, C., Eikemo, T., Van der Wel, K., & Dragano, N. (2014). A balancing act? Work-life balance, health and well-being in European welfare states. *European Journal of Public Health, 24*(3), 422–427.
Marcuse, H. (2002). *One dimensional man: Studies in the ideology of advanced industrial society*. London: Routldege.
Marx, K. (1983). Alienated labour. In E. Kamenka (Ed.), *The portable Karl Marx*. New York: Penguin.
Marx, K., & Engels, F. (1971). *The German ideology*, p. 45. Quoted in Giddens, A. *Capitalism and modern social theory*. Cambridge: Cambridge University Press, p. 63.
McKenzie, J. (2015). *Critique and happiness: Simmel, Honneth and Bauman on the contributions of social theory to a philosophy of the good life*. Paper presented at the European Sociological Association Conference, Prague, Czech Republic, 25th–28th August.
Morgan, A. (2014). The happiness turn: Axel Honneth, self-reification and sickness into health. *Subjectivity, 7*(3), 219–233.
Nettleton, S. (2013). *The sociology of health and illness* (3rd ed.). Cambridge: Polity.
ONS. (2015a). *Measuring national wellbeing: Life in the UK 2015*. London: ONS.
ONS. (2015b). *Measuring national wellbeing. Insights into loneliness, older people and wellbeing, 2015*. London: Office for National Statistics.
Ple', B. (2000). Auguste Comte on positivism and happiness. *Journal of Happiness Studies, 1*, 423–445.
Pollock, G. (2015). *The current evidence base and future needs in improving children's wellbeing across Europe: Is there a case for a comparative longitudinal survey?* (Working Paper). Dept. of Social Sciences, Manchester Metropolitan University.
Rasmussen, M., & Laumann, K. (2014). The role of exercise during adolescence on adult happiness and mood. *Leisure Studies, 33*(4), 341–356. doi:10.1080/02614367.2013.798347.
Roberts, K. (2007). Work life balance—The sources of the contemporary problem and the probable outcomes. *Employee Relations, 29*(4), 334–351.
Sayer, A. (1992). *Realism in social science: A realist approach* (2nd ed.). London: Routledge.

Sayer, A. (2011). *Why things matter to people: Social science, values and ethical life*. Cambridge: Cambridge University Press.
Sen, A. (1999). *Development as freedom*. Oxford: Oxford University Press.
Simmel, G. (1950). *The sociology of Georg Simmel*. London: Free Press, pp. 409–424.
Standing, G. (2014). *The precariat: The new dangerous class* (2nd ed.). London: Bloomsbury Press.
Taylor, C. (1991). *The malaise of modernity*. Toronto: House of Anansi Press.
Thin, N. (2012). *Social happiness: Theory into policy and practice*. Bristol: Policy Press.
Veenhoven, R. (1984). *The conditions of happiness*. Lancaster: Kluwer Academic Publishers.
Veenhoven, R. (1999). Quality of life in individualistic societies: A comparisons of 43 nations in the early 1990s. *Social Indicators Research, 48*, 157–186.
Veenhoven, R. (2004). Happiness as an aim in public policy: The greatest happiness principle. In A. Linley & S. Joseph (Eds.), *Positive psychology in practice*. New York: Wiley.
Veenhoven, R. (2008). Sociological theories of subjective wellbeing. In M. Eid & R. Larsen (Eds.), *'The science of subjective wellbeing' a tribute to Ed Diener* (pp. 44–61). New York: Guildford Publications.
Vowinckel, G. (2000). Happiness in Durkheim's sociological policy of morals. *Journal of Happiness Studies, 1*(1), 447–464.
Walker, P., & John, M. (Eds.) (2012). *From public health to wellbeing: The new driver for policy and action*. London: Palgrave.
Warren, T. (2004). Working part-time: Achieving a successful 'work-life balance'? *British Journal of Sociology, 55*(1), 99–122.
Weber, M. (1990). *The protestant ethic and the spirit of capitalism*. London: Allen Unwin.
Weber, M. (1993). *The sociology of religion*. New York: Beacon Press.
Williams, M., & Penman, D. (2011). *Mindfulness: A Practical Guide to Finding Peace in a Frantic World*. London: Piatkus.
Young, M., & Wilmott, D. (1962). *Family and kinship in East London*. Harmondsworth: Penguin.
Zingerle, A. (2000). Simmel on happiness. *Journal of Happiness Studies, 1*, 465–477.

7

Happiness and Young People

> It was the best of times, it was the worst of times, it was the age of wisdom, it was the age of foolishness, it was the epoch of belief, it was the epoch of incredulity, it was the season of Light, it was the season of Darkness, it was the spring of hope, it was the winter of despair, we had everything before us, we had nothing before us, we were all going direct to Heaven, we were all going direct the other way. (Charles Dickens '*A Tale of Two Cities*' 2003:1)

In the preceding chapters I explored some of the many different ways that writers and researchers have understood happiness. In this first of four empirical chapters I apply some of these different concepts to the lives of ordinary people—in this instance a group of five young people. I explore whether their experiences of happiness conforms to the images of wellbeing constructed by existing approaches in happiness studies. Is happiness, for example, problematic for them, as some sociologists might suggest, as their expectations of a good life are increasingly unrealistic in times of austerity? Or are young people able to experience wellbeing in more positive ways as some psychologists argue as the youth phase offers the freedoms for self-development? Young people's wellbeing is somewhat of

M. Cieslik, *The Happiness Riddle and the Quest for a Good Life*,
DOI 10.1057/978-1-137-31882-4_7

a riddle, as Dickens captured in *A Tale of Two Cities* and I use my interviewees' life stories to understand more clearly this ebb and flow of young people's happiness.

The 5 young interviewees were chosen as they were all under 25 years of age and also reflected differences in class and gender. Louise was 22 years of age when interviewed in 2012 and came from a disadvantaged background, her mother in receipt of welfare benefits. Teresa was 21 years of age at interview in 2011 and her parents were more affluent than Louise's and had skilled clerical jobs. Alice was a young black woman, aged 22 at interview, whose adopted mother was in a managerial job and hence higher socioeconomic status. The two young men who were interviewed differed in terms of their class position with John, who was 20 years of age, coming from a single parent family, his mother working in a skilled clerical job whilst Greg was 24 and his parents were better educated and held managerial positions in local firms.

One of the distinctive features of young people's lives is how they undergo a wide range of changes in a relatively short space of time—movements through education, in and out of friendship groups, exploring intimacy and sexualities as well as work identities and shifting relationship with family members. The youth literature usually characterises this as a period of 'storm and stress' (Hall 1904; Erickson 1968) and the transition to adulthood can render young people as vulnerable, at risk or as the perpetrators of trouble for others in society (Pearson 1983). It is common for youth researchers to paint the youth phase as one of many challenges with a focus on problems such as educational under-achievement, unemployment, poverty and social exclusion (Furlong and Cartmel 2007). There has been an expansion of this 'problems of youth' approach in recent years, as commentators have charted the impact of austerity on young people in Western societies (Shildrick et al. 2012; McDonald and Shildrick 2013). These analyses feed into social policy agenda that target young people's transitions and which usually suggest reform of education and training, welfare benefits and social services. Other commentators, taking a less pessimistic line, highlight the more positive aspects of youth cultures, acknowledging that the youth phase is also about fun, leisure, freedom and the important role of music, style and consumption to young people's lives (Bennett and Hodkinson 2012). Consequently, debate about young people's wellbeing reflects these divisions oscillating between doom laden pronouncements about the epidemic of ill-being amongst modern youth (Cutler et al. 2001, Wasserman et al. 2005; Robb 2007; Guardian 2014) or much more measured reports that also stress the flexibility and resilience of young people coping with the pressures of modern living (Coleman and Hagell 2007). These contradictory representations of young people echo the writings of Dickens and the paradox of youth as 'the best of times, the worst of times'. Overall, however, there has been little systematic research into the wellbeing of young people that aims to take a balanced view of their lives and which avoids these extreme positions. In contrast to these polarised approaches my interviewees' experiences showed that at times there were very real threats to their wellbeing often associated with challenging transitions events but that these times of sadness and suffering were also complemented by many other more positive events. In taking a longer biographical view we can see that for many there are ups and downs in their subjective wellbeing.

The Restructuring of Youth Transitions and the Influence on Wellbeing

All of the young people I interviewed talked about events that had been challenging for their wellbeing, and we can see that many of these reflect changes to opportunity structures, particularly in education, housing and employment. At the same time their wellbeing was also influenced by their different abilities to adapt to or cope with these structuring processes. We witness this process when we explore the experiences of Teresa who was from an upper working-class background and lived with her parents in the North East of England. The wellbeing questionnaire she completed suggested she was mostly very positive about the different aspects of her life such as home, leisure and family (scoring these 8 or 9 out of 10), suggesting she was happy 90 % of the time. Some questions she scored lower such as work and relations with nature and also suggesting that she often gets annoyed and irritable. Her experiences illustrate how some of these changes in social policy and economics can impact on the biographical routes of young people that in turn can impinge on their happiness. When asked about difficult times in her life Teresa spoke about going to university and realising that this was a mistake and having to find a job instead.

> I loved sixth form but it was just completely different at university… I didn't seem to be learning anything. It wasn't what I thought it was going to be…And there were a lot of older students; there were only ten or so young students on the course…There are no teaching jobs and I just didn't want to be stuck after three years with loads of debt and no way of getting a job. My cousins girlfriend did the same course I was doing, she just finished when I started and she's still not got a job and that was two years ago. My other friend has gone to Australia to get a job because there are no teaching jobs. (Teresa)

Although Teresa admitted there were several reasons for her decision to drop out of university she was concerned about the cost of the course fees and graduating with considerable debts, particularly when some graduates struggled to find jobs. There has been much debate and research on this issue with some suggesting that some students from poorer backgrounds are deterred from attending university because of high fees and much competition for graduate jobs (Crozier et al. 2008; Sutton Trust 2014a). Those from less affluent backgrounds, like Teresa, can suffer greater anxiety about money which can deter them from attending university or influence them to apply for more vocational courses they believe will enhance their employability (Sutton Trust 2014b; Purcell et al. 2009). Such instrumental choices, though appearing sensible, may not always lead to a fulfilling and enjoyable time at college and can lead to disenchantment and dropout (Quinn et al. 2005). For Teresa this period was a difficult time. Not least because it involved reframing her ambitions for the future but also she felt that she had let her parents down as they were both keen that she attended university.

> I was worried about my mum being angry that I wanted to leave…Because she was like, 'what are you going to do instead'? And I was like, 'I've already thought about it', and we like talked about what I wanted to do…I did think a lot about it before I told my mum to make sure it was what I wanted to do…So after university I tried to be an air hostess and then I

> started my training a couple of months later so it wasn't like I was hanging around waiting. (Teresa)

Teresa enjoyed the six months training programme in London she undertook with a well-known budget airline. Though at the end of this period she was based in Edinburgh and would often come home at weekends to visit family and friends in the North East of England as she said that she was lonely and missed her boyfriend. Here we can see the way that wellbeing is bound up with difficult choices and trade-offs, balancing the need for a job against the desire to stay near family, partners and friends. When young and inexperienced it is sometimes difficult to make these sorts of choices and anticipate the outcomes, as was the case with Teresa. The long hours at work and lengthy drives down to the North East meant that she was often over-tired and on one such journey Teresa crashed her car and spent several months recovering from back and neck injuries. The airline company, Teresa suggested, was not supportive during this period and she had to travel the 150 miles to Edinburgh every week to complete a sick note. So after a few months she resigned from the job and found herself unemployed again. Teresa spent two months unemployed and was fortunate to secure an apprenticeship position as an administrator at a local firm. At the height of recession during 2009–2010 there was much competition in the North East of England for the few available clerical jobs—though her good GCSE and A Level qualifications helped Teresa. Although she was initially relieved at finding work she spoke at length about the things that she has to manage in her new routine and the aspects that threaten her wellbeing.

> I am up at six thirty, because the traffic is always bad as well so if we leave earlier then I get to work on time…where I live the buses only come once every half hour so I would either be late or be about 45 minutes early. And getting home is a nightmare. Just being in the car in general in traffic. I get really bad road rage. It's just so expensive but is probably just as expensive getting a bus pass…loads in petrol. I hate working for nothing. At the end of the month if I've got something to pay for and I don't have enough, my mum will pay and I'll give it to her back. It just means coming off the next payday…I used to go out four times a week when in school but now it's once a week or maybe once every two weeks. (Teresa)

Teresa's case illustrates some of the characteristic ups and downs of young people's lives—the struggle to secure a job or training and associated anxiety, then relief when one succeeds in ones ambitions only to then experience disappointments as jobs and courses prove difficult or challenging. Research by psychologists has identified many of the experiences, noted by Teresa that can be corrosive of good wellbeing. Surveys have documented the way that lengthy commutes into work through busy congested streets can depress the emotions of individuals (ONS 2014). Sociologists would also point to a range of wider social developments that have created these circumstances in English cites compared to European countries. Research highlights the privatisation of public transport increasing its costs to passengers and adding to road congestion (Lucas et al. 2008; Sustrans 2012). Commentators have also documented the severe cuts to local authority funding and provision of services in areas such as the North East of England and their dire impact on the quality of life of local people (Harris 2014). These long-term policies that have evolved over many decades illustrate how historically and culturally specific structures influence many of our experiences travelling to work and in turn the quality of our lives (Bradshaw 2015). By comparison if we examine the policies in Denmark, Netherlands or Sweden and their superior transport systems we have a glimpse of how Britain might have been if governments had made different choices over the last 30 years.

A common feature of the young interviewees was their concern over money and this had a corrosive influence on their wellbeing. These findings echo those of the economists we encountered earlier on who charted the significance of good incomes and affluence for levels of happiness (Easterlin 1974; Deaton 2012). Some commentators have pointed to the 'free market' or Neo-liberal policies in the UK that have created poor quality labour market opportunities for young people which have depressed incomes which fail to meet the costs of young people's lifestyles (Howker and Malik 2010; Willetts 2011). Data suggests the relative cost of living in the UK for young people is more expensive than in other European countries. Berlin for example, despite being a vibrant capital city where 90 % of residents rent their homes is far cheaper to live (around £500 per month for a small flat) than most British cities where rent can average £700–£800 per month for a similar property (Guardian 2015). These differences stem in part from the deregulation and marketisation of rental property in the UK which has increased rents and reduced the rights of tenants compared to other countries. Teresa finds herself in the bizarre position of having to buy a car as her job is a long way from her home and the public transport is costly and infrequent. Yet the low wages she receives at work barely covers the cost of purchasing the car and running it—she has little money at the end of the month to spend. After several months of working she spoke of how the initial relief of finding work had changed to one of frustration and anger about her situation.

Teresa's experiences illustrates one of the riddles of happiness research as studies show the corrosive effects of unemployment on wellbeing (Bell and Blanchflower 2011) yet employment can bring benefits (Csikszentmihalyi and Leferve 1989) and also new threats (Erikson 1986) to happiness. The struggle to manage financially has been compounded by tensions at home between Teresa and her parents and younger sister and so in interviews she talks of now wanting a bit more freedom and space to herself. Like many young people who live at home with parents having just a small bedroom and no space to entertain friends can become, with time, a source of irritation. She suggests that she often feels 'fed up' when thinking of the future, as she would like to move out into a flat and have some independence but knows she would be unable to afford this on her low income.

MC: If you think of things you would change in your life?

Teresa: I would have more money. Definitely change that. I would move out…I just want my own house…I need my own space…but I'll need a permanent job…I've worked out I'll need £800 per month.

Teresa's responses here reflect some of the data from surveys (Easterlin 1974; Veenhoven 1999) that suggest that those on lower incomes are more likely to experience improvements in their wellbeing than more affluent individuals. This might be due to this income enabling significant improvements to lifestyles (such as independent living and establishing partnerships) that have real positive benefits to wellbeing. Although Teresa spoke of her unhappiness working long hours in a poorly paid job and her doubts about the future her overall wellbeing in recent years had been much better than some of the other young people I interviewed.

Though differences in wellbeing between the young people could be attributed to differences in material resources and opportunities it also reflected differences in the quality of relationships that young people experienced—a point that echoes some classic sociological writings that we saw earlier from Durkheim and Simmel. Those that had experienced multiple difficulties in life both materially as well in terms of social relationships tended to have had episodes of poorer wellbeing as we see with the case of Louise.

Multiple Disadvantage, Marginal Transitions and the Influence on Wellbeing

When I interviewed Louise she was 24 years of age and coming to the end of her university degree. Although she spoke positively about many aspects of her life she admitted to having recently experienced episodes of depression and had been prescribed antidepressants and counselling sessions by her doctor. Louise felt uncomfortable completing the wellbeing questionnaire so the interview focused more on her narratives of wellbeing. Louise's case illustrates the need for a complex approach to studying wellbeing (that draws on the 'sociological imagination' [Mills 1959]) to analyse the structural features of a life as well as the ways that people actively manage their wellbeing as they age and cope with various challenges and crises. Louise spoke of growing up in a poor neighbourhood in the North East of England, her parents had separated and her mother, who she lived with in a small council house, had struggled with mental health problems and drug addiction. Her two older brothers had developed their own problems with drugs and offending, spending time in prison as she was growing up. She attended local schools but her dyslexia went undiagnosed for many years so she became known for her disruptive behaviour and under-achieved academically for much of her time in education.

> I was 13, 14, mum was selling (drugs) to support us…and my mum was rattling off heroin and I was going to school at the time…I had to lock her in her room in the morning, go back at break, make sure she was alright,

> give her a drink, change the sheets if I had to, clean her up, go back to school, do dinner...And it was me in there when the house got busted, it was me getting strip searched by the police when I was young...my brother beat my mum up, grabbed by the throat and threw her around the room... he hit me over the head with a bottle...my childhood got taken away from me, I've had to grow up very quickly. (Louise)

A traditional approach of youth researchers would be to focus on the ways that job opportunities, communities and families have changed, creating patterns of social disadvantage that have shaped Louise's life chances and her poor wellbeing. Her story is a familiar one that we have seen in other societies that have suffered de-industrialisation and effects of Neo-liberal reforms of welfare such as the American cities of Detroit, Baltimore and Cleveland (Sennett 1998). As I have written elsewhere (Cieslik and Simpson 2015), the marketisation of welfare provision and educational opportunities in England have also adversely affected the lives of many working-class young people. In recent years, as many researchers have documented (Aldridge et al. 2012; Clarke 2014), there has been a cumulative process whereby multiple disadvantages in many different areas of life combine to undermine the wellbeing of the poor in Britain. For Louise her sense of self and quality of life were influenced by growing up in a fragile family, with little money, in poor quality housing and studying at a poorly performing school.

In the interview Louise and I generated a biographical timeline where we charted these different aspects of her biography and the life course transitions she made. Bourdieu's analytical framework (of habitus, capitals and field) allows us to disentangle how a paucity of resources influenced the structuring of disadvantage and wellbeing so that Louise followed a predictable path from home, through education and into early adulthood. Poor material resources because of family breakdown and unemployed status of Louise's mother meant that Louise had no option but to rely on state housing which in turn meant that her educational opportunities were also limited. England like the USA has become a society marked by significant inequalities in educational provision—with outstanding schools serving affluent neighbourhoods existing a few miles from struggling schools that cater to children of poorer districts

(Wilkinson and Pickett 2009). Unlike more affluent children, there was little money to pay for travel to another school or for specialist tutors. Louise spoke of not fitting in at school and not enjoying her time in education because of her dyslexia. Some of this poor wellbeing could have been avoided if there had been resources to pay for additional help or a move to a different school—the sorts of advantages enjoyed by more affluent children as their parents use their material and social capital to purchase cultural opportunities for them (Power et al. 2003). Much research into inequality in the UK has documented how the disadvantaged like Louise are unable to compete effectively in a market-based system of schooling, housing and welfare (Dorling 2011). This has a corrosive effect on their wellbeing as they watch their richer peers succeed whilst they miss out on opportunities and chances to develop their potential. This is one of the devastating effects, according to Bourdieu, of the way that social fields operate to reproduce inequalities and symbolic violence (Bourdieu et al. 1999).

Success and Happiness Against the Odds?

As I have suggested, one of the riddles of young people's wellbeing is that many surveys point to widespread experiences of anxiety and uncertainty yet at the same time the majority of young people state in these surveys they are happy most of the time (UNICEF 2007; Oswald and Blanchflower 1997). We can see evidence of these contradictions in our interviewees' lives where shifts in life course transitions can engender negative emotions yet they can also have cultural lives that bolster a positive sense of self. In Louise's case we see a similar complexity. Louise's early experiences of social disadvantage would suggest to many that she would grow up to repeat the marginal transitions and poor wellbeing that her mother had endured. During the interview she spoke repeatedly of her fear of ending up like her mother—few qualifications, unemployed and problems with drink and drugs. Survey data and sociological research over many years have indeed documented these inter-generational relationships between socioeconomic background and patterns of attainment at school, entry to university and into employment (Sutton Trust 2014a, b; Bourdieu and Passeron 1990; Jones 2002). Despite the barriers that working class youth have to overcome, studies suggest that many do well in education and into work, and this was the case with Louise. In interview discussions we explored how she had managed to pass her school examinations and gain entry to university despite all the difficulties of her childhood.

> I think the massive change for me breaking the cycle, is that I found roller hockey…We had no money but mum saved up for Christmas and I got some skates and a stick and it progressed from there and I ended up getting into a proper team…instead of me going and getting drunk and taking drugs with me mates I used to go training or have games on a weekend and so it gave me a structure. My mum always paid for me to go to work and it used to cost her more to pay for me to get a lift to work and back than my wages but she said I needed to do it to show that it's the right way to live your life, by working and that's what she did and I think that's what broke the cycle for me. (Louise)

Other researchers have noted the important role that sport can play in improving the wellbeing of young people (Rasmussen and Laumann 2014) and with Louise it does seem that her sporting interests allowed her to develop more positive orientations to life and schoolwork in particular. Given the extent of her problems, however, Louise acknowledged that there were other aspects of her life that helped her to overcome some of the difficulties she faced. One resource that was significant was the extensive friendship network that Louise enjoyed during her adolescence—she said that she never had problems making friends and these helped her to manage many of the emotional problems she encountered growing up—without these her wellbeing would surely have been much worse (Griffin et al. 2009). Louise also spoke about the important role of her grandparents during her youth and they undertook many of the nurturing roles that parents would normally perform. As is the case with many families it is often the grandparents that offer children a predictable, constant source of practical and emotional support aiding transitions and also wellbeing during adolescence (Chan and Boliver 2013).

Success Yet Poor Wellbeing: The Shadows of the Past

Despite Louise appearing to have escaped her past life of disadvantage and achieving the successes that are usually associated with a happy life—university degree, a job, partner, her own home and circle of friends—she acknowledged that her life had not been that straightforward. Here we see how there can be dissonance between the objective situation of individuals and their subjective wellbeing. To the casual observer Louise's successes in adult life might also suggest improvements in her wellbeing from when she was young but our interviews together suggested a complicated story. For several years she had experienced bouts of anxiety, depression and anger at the way her life had worked out. In her final year of university Louise's mother was diagnosed with incurable emphysema and this, she says, triggered a lengthy period of depression.

> I'm on tablets now for depression…I went on them last year when my mum was ill…If I wasn't crying I was just shouting for no reason and I was just a proper mess…after counselling and stuff being able to recognise the emotions I was actually having and what they really meant, because I thought that I was sad, I was scared of losing my mum, but I wasn't. I was angry.

Louise discusses an issue that we see in many other interviewees—how there is suffering and sadness despite the external markers of achievement and what appear to be successful life course transitions. It is this sort of complex and ambivalent aspect of wellbeing that quantitative surveys struggle to capture with their simple questions. How might we account for this dissonance between objective circumstances and subjective state? Some writers in the past, drawing on psychoanalytical ideas, have suggested that early childhood events can be internalised and leave a lasting psychic legacy that can be corrosive of wellbeing (Craib 2001; Sennett and Cobb 1977). These arguments about the 'psychological aspects of oppression' or the 'hidden injuries of class' suggest that early classed experiences (such as attending 'poor' schools and living in deprived neighbourhoods) can create a lifelong sense of inferiority (Cieslik and Simpson 2015). Bourdieu also deploys this sort of analysis in his work on class habitus (Bourdieu and Passeron 1990), particularly how working-class students are treated in schools, as we see in Louise's case, and the impact this has on her sense of self (Bourdieu and Passeron 1990; Bourdieu et al. 1999). The significance of habitus is how the everyday experiences of social exclusion and class domination come to be internalised by people like Louise, creating patterns of thought, ideas and values which are disabling and threaten wellbeing which in turn can further add to exclusion and domination.

Yet when we read Louise's testimony there are a multiplicity of events that have contributed to these hidden injuries—many of which extend beyond social class—and so there is a need for a wider reading of Louise's experience and a more general psychoanalytical approach to her biography.

> When I was growing up I felt like I had lost my mum because of her drug addiction…Me and my mum can openly talk about things and she said to me, 'I'm not going to lie to you Louise, but if I had not taken drugs I would have killed myself'. But I said, 'you don't understand, that to me you were dead because you weren't yourself'. Then she finally gets off them, battling addiction and I get her back and then she gets emphysema and she's dying…I was angry at my mum, 'how dare you get ill', and I was angry at the world because I'm losing my mum and there's nothing I can do about it. (Louise)

The anger she feels which she says has been corrosive of her wellbeing in recent months stems from the many different ways she supported her mother when she was a child—all that she gave up and lost in her childhood. And now it seems, with the illness of her mother all that work to save her mother from a life of drugs has been for nought. When looking back on recent years, though Louise was at times happy and contented she also had episodes of stress and anxiety that culminated more recently with a prolonged period of depression. She talks repeatedly of 'coping' with her difficult childhood by 'putting a brave face on things' but that at the same time she had also felt very angry at seemingly unrelated events. Like many people today Louise found herself at the mercy of thoughts and emotions that seem to be out of control and often disconnected from our efforts to manage our wellbeing. This became very troubling and led her to her doctor and a course of counselling. Over a period of several months Louise was helped through counselling to unpack these emotions drawing on some of the ideas from psychoanalytical theory. She suggested that the counselling helped her to realise that her blocking out her desires for a 'normal life' and her frustrations at her life had become internalised over many years and these powerful feelings had at times leaked out into her everyday life causing anxiety and stress.

Psychoanalytical theory tells us that we manage the ups and downs of everyday life by repressing those feelings that are threatening but that such repression is only ever partial and can seep out into our conscious life and again threaten us (Elliott 2015: 51). We are however often unaware of this process and so are confounded when we experience emotions that are seemingly unconnected to a conscious event. This notion of repression and the way the unconscious influences daily life is one of the reasons why happiness is such a riddle. For we often plan out our lives to be happy and successful unaware of how there are deeper parts of the self that have been shaped by our earlier experiences that also influence our wellbeing. Louise's case does suggest that there are limits to more traditional social science accounts of wellbeing that use simple questions about life satisfaction or emotions. This ebb and flow of wellbeing also seems far removed from the 'broad brush' arguments offered by some commentators who view happiness as a social problem today (Furedi 2004; Ahmed 2010).

There were other examples of young people who had made important life decisions, trying to flourish in the face of life's challenges and adversity, only to find themselves surprised at their emotional responses and unsure how to interpret what had happened to them and how to proceed in life.

Alice's Story

At interview Alice was a 22-year-old university student of African-Caribbean heritage who had grown up in South London. She never knew her father, living initially with her mother and then a family friend (who she referred to as her aunt) when her mother died when she was 13 years old. The questionnaire that Alice completed suggested that overall she tends to be 'pretty happy with life at present', around 45 % of the time (35 % unhappy and 20 % neutral). For some other questions around her studies and leisure she scored 7 out of 10 and scored less for family, exercise, home and main relationship (3–5 out of 10). The happiness mapping exercise we conducted revealed a pattern of good and bad experiences and emotions during her life and which, as the survey results suggested, still shaped her wellbeing. There were a number of things, Alice suggested, that were difficult to deal with when growing up—in a single parent family money was tight, so often Alice felt that she could not afford the lifestyles that some of her friends enjoyed—nights out, expensive clothes and regular holidays. This account conforms to other research that points to the significance of a good family income for wellbeing (WHO 2009) and in particular for young people's wellbeing that involve consumerist lifestyles (Deutsch and Theodorou 2010). This illustrates one of the arguments proposed by sociologists that popular ideas of happiness can be oppressive as they encourage superficial, consumerist and often unobtainable ways of living, that ultimately offer unsatisfying experiences of happiness (Marcuse 2002; Furedi 2004). Losing her mother and not knowing her father meant that Alice felt different from her friends who all lived with family members—this was a source of sadness and one that she tried to cope with as best she could. As a young black women growing up in a working-class district of London she had experienced racist abuse and there had been some tensions between different peer groups at school (see Webster 2009). These challenges to her wellbeing were countered by the support she received from her friends at school—she mixed with other black students so they could manage some of the exclusion they experienced as minority pupils. Alice also spoke about working hard at her studies, 'being a geek' as a way of getting on in life and being positive about the future—an aspect of cultural identity and a 'coping strategy' that has been noted elsewhere (Reay 2001). These positive attitudes were also reinforced by her faith as a Christian which she said was important by offering a set of principles to live by as well as a community of friends that she met regularly at church activities (see Layard 2005: 72). These sorts of experiences may account for some of the diversity we see in the wellbeing surveys conducted on young people—episodes of anxiety and bad feeling whilst also other more positive events that support good wellbeing.

Life's Full of Surprises: Alice and the 'Gap Year'

One theme of this book is how we are all fallible—that we often assume we know ourselves (and those around us) better than we actually do (see Kahneman 2011). There are parts of our internal lives that we neglect and hence we often make poor choices about what we want in life and what it is that makes us happy (Gilbert 2006). We witnessed an example of this with Louise and how some of her childhood experiences resurfaced as an adult, undermining her wellbeing in the process. Alice too offered another example of how traumatic childhood events can be internalised only for these emotions to re-emerge in unexpected ways as an adult. Alice felt that working with young people would be a good career so took the opportunity to have some work experience in this area during her gap year before starting university. She used her contacts with her local church to secure a place as a youth leader in a small church and lived with the church Elder and his family in the North East of England.

> I've achieved a lot [on the gap year], I was on my own, people I hadn't met before, I was settling in. You know, I've really become an adult…It also helped me to see how much I enjoy working with kids…It felt like the right thing to do…There's this verse in the bible, 'wherever you go I'll be with you, just be strong, be brave'. (Alice)

At first Alice enjoyed her time away and the work with the children at the different youth clubs she helped to run. But within a few months she began to feel uneasy about her life with the church leader and his family and she spoke of feeling cross and angry about the routines of her life in the North East. During the interview Alice suggested, just as we saw with Louise's accounts, that she was really surprised at these negative emotions that seemed inexplicable having emerged from what seemed innocuous, everyday events. It just seemed so irrational to feel angry towards the church leader, his wife and children who had taken her into their home and helped her with the work experience that was furthering her studies and career.

> I hadn't lived with a family in a very long time. So I live with my aunt because my mum died when I was younger, so it's always really been me and my aunt. So I think to come into a situation where I was living with a family—I think it was just a bit weird for me. I don't know, I felt like—I think it really forced a sense of not-belonging, that I'd had that I didn't realise I had…And they tried as much as they could to include me and stuff, but I just—I don't know, I think it brought up things, like feelings maybe I didn't realise I had, at the time. I think about family, and I think about belonging as well. Because out of my friends that are closest to me, I'm the only one that has a single parent, and isn't even her parent…I had a bit of a bad year, I was feeling really down and I really lost where I was, I think. So after I was like, 'Well, maybe God doesn't exist, because why would God tell me to come here?' So it wasn't a happy time at all. (Alice)

For several months then, Alice felt unhappy whilst on her gap year, despite her expectations that it would be something that she would enjoy and would be useful for her career. She was miserable, not just because of a sense of unease about living in an, 'ideal family' but also that these negative feelings seemed inexplicable at first, having come 'out of the blue'. As Alice had felt that the decision to do a gap year was a considered one, informed by her faith, these negative events also challenged her religious beliefs and values. This she took very seriously as her faith was at the core of her identity, guiding her in her everyday life. Alice spoke of how for a time she experienced a crisis of identity, one that profoundly threatened her wellbeing so that she felt ill at ease and anxious for several weeks.

As with other examples we have seen, this account from Alice illustrates one of the riddles of happiness. That we often push ourselves to do things that we feel are good for us; will help us to succeed in life or that we think may make us happy only to experience unexpected negative emotions. The inexplicable nature of these feelings can be unsettling and can further threaten our wellbeing for we realise that thinking them through rationally often fails to render them intelligible. For as we struggle to rationalise about where these feelings may come from we are faced with the reality that emotional responses to events are often triggered by seemingly unconnected life events that defy rationalisation. As we did with Louise earlier we can use some psychoanalytical concepts to offer an interpretation of Alice's experiences. With Alice's case it seems that the gap year experience of family life unexpectedly triggered deeply held emotions about her own experiences of losing her mother and father and being different to other children which in turn threatened her deeply held religious beliefs. When young, Alice may have 'managed' some difficult events by internalising or repressing the associated emotions that can then influence wellbeing later on in life. Taken all together there were a confluence of negative emotions and events that challenged her wellbeing. And it was only after she had finished her gap year and started her university degree that she was able to make sense of this turbulent time in her life. As I discuss later on, one of the paradoxes of happiness is that amidst what appear to be difficult times there can also be 'silver linings'. For Alice, 'her gap year that went wrong' revealed aspects of herself that she was unaware of and so with time these difficult events also had a positive side as it allowed Alice to come to understand herself a little more as she made the transition to adult independence. She realised there were powerful emotions about her family—having lost her mother and not knowing her father. Out of this difficult time came the decision to track down her father in Jamaica which she hoped would help her make sense of her background and her identity (Craib 2001).

Happiness, Fateful Moments and Hidden Resources

When we turn to the young men who were interviewed we can see some different experiences of happiness compared to the women. In particular, Greg's story is interesting as it illustrates the significance of a key life event and its impact on wellbeing. He was 24 years of age when first interviewed in 2011 and scored himself lowly in the wellbeing questionnaires he completed. He scored 5 or 6 (out of 10) on many of the scales that measured satisfaction with different areas of life (when many of the others in the sample scored 7 and above), saying that he was only happy about 10 % of the time. The self-report questions also suggested he had little optimism about the future, and was pessimistic about many different aspects of his life such as leisure, health, control, finding meaning in life. His story also contrasts markedly with those of Alice and Louise. Where for the young women problematic family events can shape 'hidden injuries', for Greg early positive experiences of family relationship offered resources that were useful for coping with life's challenges.

At the interview Greg spoke very positively about his childhood in the North East of England. His parents had a loving relationship; he and his sister were nurtured and they both did well at school and made friends and had good social lives as young adolescents. The only issues which Greg spoke about as being a challenge for his wellbeing when young were those associated with his tendency for obsessive behaviour. This was something he felt might be a family trait as his sister had obsessed about her weight and developed anorexia when a teenager. After gaining his A-level qualifications Greg attended a local university studying for a computing degree. It was at this point in his life that Greg spoke about developing an interest in bodybuilding. There were lots of positive effects of such an interest he suggested—regular workouts at the gym, eating sensibly, learning new skills, insights into physiology and diet and making new friends. These accounts echo the writings of Csikszentmihalyi (2002) and his discussion of 'flow' experiences where one can lose oneself in an activity as long as it is challenging and needs some concentration—physical workouts are a classic example of these. There was a sense of satisfaction in living a more disciplined life and then witnessing the effects of this working on the body—putting on muscle mass and developing good tone. These views reflect research in youth studies that point to young people's views about the greater control they have over their identities and lifestyles than those in the past—aided in part by the expansion of leisure opportunities and greater freedoms of self-expression (Miles 2000).

In the interview, Greg showed me photos of his body before and after his body building regime and how marked the changes were. Overall Greg felt that these few years at university when he was bodybuilding were happy times—a real positive for him was the sense that many of the women he met (in his part-time job as a barman) were also impressed with his muscled physique. He felt that his work on his body had enhanced his attractiveness to women and as a result he felt more confident about talking to women and asking women for a date. Greg did admit, however, making the connection to his sister's experiences, that he had become aware of the dangers of taking bodybuilding too far—becoming obsessed with his appearance and training regimes so that it undermined his wellbeing. This darker side of lifestyle choices has been noted by others where the freedom to explore ways of living can escalate into obsessive routines that can undermine sociability and create insecurities (Sweetman 1999; Frost 2001; Furedi 2004). As it transpired, Greg never had the opportunity to take his bodybuilding too far as a fateful moment intervened, transforming his life.

A few months before the end of his degree programme Greg was out drinking with a few friends when he was involved in an accident, falling out of a moving taxi, sustaining serious head injuries. Greg spent several months in hospital and had 18 months of further rehabilitation by the time of the first interview. He had been left with partial paralysis in his left leg, hip and arm and a range of cognitive difficulties—his memory had been affected and speech and general processing now took much more effort and so he was much more fatigued than before the accident.

Greg suggested that before his accident he was quite content overall with his life as he was finishing his degree, considering graduate study, worked part-time job, shared a flat with some friends and had a good social life. Research by others (Robb 2007) point to the significance of these social networks, good family relationships, education and employment for good wellbeing in young people. Yet all of these features of Greg's life were transformed by his accident and the subsequent upheaval presented many challenges for him.

> I remember the first thing when I woke up from the coma, I said, what happened, why am I here, what's wrong with my leg, what's wrong with my arm and that's what I was bothered about…I couldn't speak at all, like if someone asked us, 'do you want a pot of tea?' I'd just say silly stuff, 'Oh yes I'll have a kettle please'…It's annoying because I know how much I used to go to the gym and it just takes your mind off everything, but then what can you take your mind off when you don't have the gym. (Greg)

I first met Greg at a support group run by a brain injury charity. I too had suffered a brain injury (a cerebral haemorrhage) a year before Greg so we shared stories about the problems we faced trying to lead a normal life again. These months after the accident Greg was preoccupied with physical therapy, trying to restore mobility to the left side of his body. He also underwent psychological tests and support to help recover some of his speech and memory. Although he made much progress, learnt to walk without a stick and improved his speech and memory, he realised that his old life was gone forever. His partial paralysis meant that his mobility and independence were greatly reduced. Working in a bar, nights out around the town, sessions in the gym and driving a car would all now be beyond him. We can understand now why Greg gave such pessimistic answers when completing the wellbeing questionnaire. Greg said that 'he rarely felt happy' and 'was often dissatisfied with life' which reflects how his accident had transformed his life. Eighteen months after his brain injury he is still coming to terms with the loss of what he saw as a good life. When looking towards the future he also felt pessimistic about his prospects that reinforced his pessimism.

> [The future] that's uncertain more than anything. Like I keep thinking, will I ever get a wife? I don't know if I can. You know I don't meet so many people like I used to meet, because you don't connect with people so much. It's hard to connect with people you don't see very often.

> When I'm just walking through town I basically think everyone's in darkness and me, a bright light is shining on us so everyone is looking at us and that's again what I hate. I don't mind if they're looking at us because they think, 'Ah he's nice…I just think they're just looking at my leg, they're just looking at my arm and they're just looking at the way I walk. (Greg)

Greg was acutely aware that the good things about his old life such as his friends, work, gym and girlfriend had all been taken away by his accident. The sudden collapse in his friendships and social networks had left him isolated and lonely—as we have seen these social contacts are significant for good wellbeing (Pahl 2007). Greg also suggested that he had experienced some psychological problems adjusting to his new identity as a disabled young man. At times he was able to be positive, reflecting on the progress he has made in rehabilitation and the fact that his accident could have been much worse but then at other times feeling consumed with frustration as his disabilities. He was also conscious of his new identity as disabled—sensitive to how others reacted to his disability. This stigma compounded the daily struggles he had navigating his way around (Goffman 1963). And so he often feels angry at his misfortune—why had he been drunk, how had he slipped out of the car and why didn't the driver stop sooner? Greg's gradual efforts to rebuild his life, taking each day as it comes illustrates how for him his wellbeing is a daily effort of working at each day, making small choices and little steps towards a better future. In many ways this gradualist understanding of the process of wellbeing that we see here with Greg resembles the notion of Eudaimonic happiness depicted by Aristotle (2009) or Buddhist ideas of the journey of wellbeing (Schooch 2007: 102). It is something that one works at, overcoming challenges along the way—a far cry from popular ideas of happiness that have been used by some academic commentators (Ahmed 2010; Furedi 2004; Davies 2015) that present it as subjective, personal good feeling.

Greg's story is a classic example of how a major life event can have serious impact on wellbeing. Many surveys point to the role of good health to satisfaction with life (Children's Society 2015) and how the disabled tend to have poorer wellbeing and mental health than the general population. Yet at the same time Greg discussed how his life and wellbeing would have been much worse without a range of health and welfare services provided by the state and also the support of his parents and sister. In contrast to Alice and Louise, whose family experiences had been a source of hidden injuries, for Greg his more positive family relationships proved to be an important source of support and resilience. Greg's family had been at his bedside during his long stay in hospital and then cared for him at home, adapting their house to his mobility needs. They also helped him apply for various support grants and secure specialist rehabilitation services to aid his recovery. They also encouraged him to attend a support group, learn to drive again and begin to rebuild his friendship groups. At time of interview they had also began to help Greg find a specially adapted flat so that he could live independently again. Although Greg felt that he was dependent on his parents, at the same time they were working in various ways to help him rediscover some of the independence he had before his accident. As other research has shown (Layard and Dunn 2009) Greg was fortunate to have this support and these family and state welfare resources went some way in mitigating some of the worse effects of his accident on his wellbeing.

The 'Ebb and Flow' of Young People's Wellbeing

Young people in the UK face a wide range of difficulties during their transitions to adulthood such as the costs of education and accommodation, poor quality training and employment opportunities. This structuring of youth transitions is reflected in survey results that suggest that significant numbers in the UK suffer from anxiety and depression (Children's Society 2015; Wasserman et al. 2005; Cutler et al. 2001). Paradoxically, as we have discussed, research also points to how the majority of young people, when asked, tend to suggest they are mostly happy with their lives (Eckersley 2011: 628; Oswald and Blanchflower 1997; ONS 2015a; Layard and Dunn 2009). When compared to other age groups the young also tend to have better life satisfaction than older people (Blanchflower and Oswald 2008). How then can we account for these contradictory findings that suggest this time in life can be the, 'best of times and the worst of times'? One possible interpretation is that young people's wellbeing oscillates over time during this period of transition to adulthood. When young people encounter difficulties wellbeing deteriorates but with time can improve as many have social networks and enjoy activities that support their wellbeing. One distinctive feature of the young people in this research was the amount of free time they enjoyed compared to other older respondents. They tended to work fewer hours than older interviewees, were more likely to be single and had fewer caring responsibilities. Is there something about this free time and being 'carefree' that accounts for why young people tend to be happier than older people? Is there something about being young and having time to experiment with life that allows for hope and optimism about the future that can help support wellbeing even when times are difficult? All of the young sample members enjoyed enough free time to experiment with new hobbies and interests that may account for their relatively good wellbeing compared to older interviewees. Louise had tried out different sports at university, Greg had become interested in bodybuilding, and as we see in the following paragraphs, Alice had further developed her interest in music.

> I absolutely love playing the piano, I've been playing it now for maybe 11 years … someone from the church offered to pay for my lessons, he got me a piano and paid for lesson for 5 years until I was 18 … I like playing the piano as it makes me feel really happy. I feel like I have achieved something if I have composed a song. I feel like it's an event as well if I feeling a bit frustrated … I feel like I, know it sounds weird, but met the feeling through what I've played. It's really important to me because, I wouldn't say it was spiritual but it's quite an emotional thing, music. I remember one time, I was feeling really down and I got my guitar, played a few chords and I was like, 'Yeh', it wasn't even a sad song I was playing, I don't know whether it's a form of distraction or something. (Alice)

Just as with playing a sport, there was something about playing an instrument that allowed people to escape some of the daily negative routines and experiences that could undermine their wellbeing (McDonald et al 2012). Once again, as we discussed earlier with Greg's interest in bodybuilding, we can see the importance of flow experiences producing positive emotions that can bolster young people's wellbeing. Being able to lose oneself in something—to transport oneself somewhere else is an important feature of positive wellbeing activities. Playing an instrument as Alice recalls is significant as it involves the long-term discipline of learning and skill development, the opportunity for self-expression, creativity and the sense of achievement—all of which bolsters a sense of fulfilment and a range of positive emotions.

John, the final member of the young people's sample, illustrates this 'ebb and flow of wellbeing' particularly well. He spoke about the last few years as being a turbulent time, oscillating between highs and lows. He had achieved well at school and stayed on to study A levels but had then become disenchanted with his subjects and left after one year of the course with passes in three subjects at AS level. When he dropped out of his course he spoke of feeling particularly 'low', he had little money as he was unemployed for several months—his mother who was a single parent also struggled to support him financially. This echoes the research that documents the corrosive effects of unemployment on young people (Webster et al. 2004) and in particular its impact on their wellbeing. He was also disappointed with his education as he hoped it would lead to university and a well-paid career. When I asked John how he managed during this time he sketched out several aspects of his regular routines that had 'cheered him up'—such as time with his girlfriend, his weekly game of football with friends and the Friday night/Saturday night tours around the local pubs and clubs.

> I'm out nearly every night with friends or family … I still go out now with friends that I've known since infant and junior school … We'll go to the snooker hall, maybe spend a tenner or something… we go to the local pub for games of pool and then weekends go there before we go out to the clubs … We are bit of a critical bunch, so there's pretty much no limits to how far taking the piss out of someone can go really … we know if someone steps over the line… we don't actually mean it, you know. Whatever happens we would always be there for our mates … we're just that sort of bunch. That's our sense of humour, it's the way that we all get along. (John)

All of the five young people I interviewed said they 'felt better about themselves' if they could meet up for coffee or a beer with friends and chat and 'have a laugh'. Good friendships can be therapeutic (Pahl 2007), helping these young people to manage some of the difficult and tedious aspects of their lives—the boring jobs, problems with parents and partners and coping with little money in a consumerist society. As other researchers have noted (Chatterton and Hollands 2001) one appeal of 'a good night out' is the journey through the familiar places of one's local town, visiting pubs and clubs that have featured in countless other nights out as one has grown up with these friends. Feeling embedded and part of a place is crucial for a sense of belonging and identity—of being a Geordie, Mancunian or Scouser. The young interviewees spoke at length about the importance of their friends—there are shared traditions and rituals that characterise a night out—where to meet up, what to drink, buying rounds of drinks, the sequence of venues visited, where one sits at each place and the old stories that make up the chat and gossip along the way (Griffin et al. 2009). This is a stark reminder then of the social nature of happiness—though experienced individually at an emotional level it is also rooted in everyday sociability—a happiness that is reciprocal and relational.

In this chapter we have seen the different influences on young people's wellbeing, such as job opportunities, income and good family support. Although affluence (and social class resources) eases the transition to adulthood their impact on happiness is more complex. For all the interviewees went through some difficult times that threatened their wellbeing irrespective of their socioeconomic backgrounds. Yet the youth phase is distinctive in offering time and freedoms to flourish that are often lacking in later life and so this biographical analysis illustrated some of the ebbs and flows of young people's wellbeing. One feature of this wellbeing was the sense of hope and optimism about the future—when young there is always time to try something different or make amends. We now turn to the 'Thirty Somethings' sample and examine whether ten years later people still feel as optimistic about their lives as they did when they were young.

Bibliography

Ahmed, S. (2010). *The promise of happiness*. London: Duke University Press.

Aldridge, H., Kenway, P., MacIness, T., & Parekh, A. (2012). *Monitoring poverty and social exclusion 2012*. York: Joseph Rowntree Foundation.

Aristotle. (2009). *Nicomachean ethics*. Oxford: Oxford University Press.

Bell, D., & Blanchflower, D. (2011). Young people and the great recession. *Oxford Review of Economic Policy, 27*(2), 241–267.

Bennett, A., & Hodkinson, P. (2012). *Aging and youth cultures: Music, style and subcultures*. London: Berg.

Blanchflower, D., & Oswald, A. (2008). Is wellbeing U shaped over the life cycle? *Social Science and Medicine, 66*, 1733–1749.

Bourdieu, P., & Passeron, J. C. (1990). *Reproduction in education, society and culture*. London: Sage Press.

Bourdieu, P., Accardo, A., Balazs, G., Beaud, S., Bonvin, F., Bourdieu, E., Bourgois, P. et al.(1999). *The weight of the world: Social suffering in contemporary society*. Stanford, CA: Stanford University Press.

Chan, T. W., & Boliver, V. (2013). The grandparent effect in social mobility evidence from British birth cohort studies. *American Sociological Review, 78*(4), 662–678.

Chatterton, P., & Hollands, R. (2001). *'Changing our toon': Youth, nightlife and urban change in Newcastle*. Newcastle: Newcastle University Press.
Children's Society. (2015). *The good childhood report 2015*. London: Children's Society.
Cieslik, M., & Simpson, D. (2015). Basic skills, literacy practices and the hidden injuries of class. *Sociological Research Online, 20*(1). Retrieved from http://www:socresonline.org.uk/20/1/7.html
Coleman, J., & Hagell, A. (Eds.) (2007). *Adolescence, risk and resilience: Against the odds*. London: Wiley.
Craib, I. (2001). *Psychoanalysis: A critical introduction*. Cambridge: Polity Press.
Crozier, G., Reay, D., & Clayton, J. (2008). Different strokes for different folks: Diverse students in diverse institutions. *Research Papers in Education, 23*(2), 167–177.
Csikszentmihalyi, M., & Leferve, J. (1989). Optimal experience at work and leisure. *Journal of Personality and Social Psychology, 56*(5), 815–822.
Csikszentmihalyi, M. (2002). *Flow: The Psychology of Happiness*. London: Rider Press.
Cutler, D., Glaeser, E., & Norberg, K. (2001). *Explaining the rise in youth suicide* (Harvard Institute of Economic Research Paper No. 1917).
Davies, W. (2015). *The happiness industry: How the government and big business sold us wellbeing*. London: Verso.
Deaton, A. (2012). The financial crisis and the wellbeing of America. *National Bureau of Economic Research,* 343–368. Retrieved from http://www.nber.org/chapters/c12447
Deutsch, N., & Theodorou, E. (2010). Aspiring, consuming, becoming: Youth identity in a culture of consumption. *Youth and Society, 42*(2), 229–254.
Dickens, C. (2003). *A Tale of Two Cities*. London: Penguin.
Dorling, D. (2011). *Injustice: Why social inequality persists*. Bristol: Policy Press.
Easterlin, R. (1974). Does economic growth improve the human lot? Some empirical evidence. In P. David & M. Reder (Eds.), *Nations and households in economic growth*. New York: Academic Press.
Eckersley, R. (2011). A new narrative of young people's health and wellbeing. *Journal of Youth Studies, 14*(5), 627–638.
Elliott, A. (2015). *Psychoanalytic theory: An introduction* (3rd ed.). London: Palgrave.
Erickson, E. (1968). *Identity: Youth and crisis*. New York: Norton.
Erikson, K. (1986, February). 'On work and alienation', Presidential address, American Sociological Association. *American Sociological Review, 51*, 1–8.

Frost, L. (2001). *Young women and the body: A feminist sociology*. London: Palgrave.

Furedi, F. (2004). *Therapy culture: Cultivating vulnerability in an uncertain age*. London: Routledge.

Furlong, A., & Cartmel, F. (2007). *Young people and social change*. Buckingham: Open University Press.

Gilbert, D. (2006). *Stumbling on happiness*. London: Harper Perennial.

Goffman, E. (1963). *Stigma: Notes on the management of spoiled identity*. London: Simon and Schuster.

Griffin, C., Bengry-Howell, A., Hackley, C., Mistral, W., & Szmigin, I. (2009). 'Every time I do it I absolutely annihilate myself': Loss of (self) consciousness and loss of memory in young people's drinking narratives. *Sociology, 43*(3), 457–476.

Guardian. (2014, October 14). *Social care is on the cusp of a crisis*. Retrieved from http://www.theguardian.com/social-care-network/2015/oct/14/social-care-cusp-crisis

Hall, G. S. (1904). *Adolescence: Its psychology and its relation to physiology, anthropology, sociology, sex, crime, religion and education* (Vol. 2). New York: D Appleton and Co.

Harris, J. (2014, November 25). Mission impossible. *Guardian Newspaper*, 27–29.

Howker, E., & Malik, S. (2010). *Jilted generation: How Britain has bankrupted its youth*. London: Icon Books.

Jones, G. (2002). *The youth divide: Diverging paths to adulthood*. York: Joseph Rowntree Foundation.

Kahneman, D. (2011). *Thinking fast and slow*. London: Penguin Books.

Layard, R. (2005). *Happiness: Lessons from a new science*. London: Penguin.

Layard, R., & Dunn, J. (2009). *A good childhood: Searching for values in a competitive age*. London: Penguin.

Lucas, K., Tyler, S., & Christodoulou, G. (2008). *The value of new transport in deprived areas: Who benefits, how and why?* York: Joseph Rowntree Foundation.

Marcuse, H. (2002). *One dimensional man: Studies in the ideology of advanced industrial society*. London: Routldege.

Miles, S. (2000). *Youth lifestyles in a changing world*. Buckingham: Open University Press.

Mills, C. W. (1959). *The sociological imagination*. Oxford: Oxford University Press.

ONS. (2014). *Commuting and personal well-being*. London: ONS.

ONS. (2015a). *Measuring national wellbeing: Life in the UK 2015*. London: ONS.

ONS. (2015b). *Measuring national wellbeing. Insights into loneliness, older people and wellbeing, 2015*. London: Office for National Statistics.

Oswald, A., & Blanchflower, D. (1997). *The rising wellbeing of the young, Working Paper No. 6102*. Cambridge, MA: National Bureau of Economic Research.

Pahl, R. (2007). Friendship, trust and mutuality. In J. Haworth & G. Hart (Eds.), *Wellbeing: Individual, community and social perspectives* (pp. 256–270). London: Palgrave.

Pearson, G. (1983). *Hooligan: A history of respectable fears*. London: Macmillan.

Power, S., Edwards, T., Wigfall, V., & Whitty, G. (2003). *Education and the middle class*. Milton Keynes: Open University Press.

Purcell, K., Elias, P., Atfield, G., Behle, H., Ellison, R., with Hughes, C., Livanos, I., & Tzanakou, C. (2009). *Plans, aspirations and realities: Taking stock of higher education and career choices one year on*. Coventry: Higher Education Career Services Unit (HECSU) and Warwick Institute for Employment Research.

Quinn, J., Thomas, L., Slack, K., Casey, L., Theaton, W., & Noble, J. (2005). *From life crisis to lifelong learning. Rethinking working-class 'drop-out' from higher education*. York: Joseph Rowntree Foundation.

Rasmussen, M., & Laumann, K. (2014). The role of exercise during adolescence on adult happiness and mood. *Leisure Studies, 33*(4), 341–356. doi:10.1080/02614367.2013.798347.

Reay, D. (2001). Spice girls, 'nice girls', 'girlies' and tomboys: Gender discourses, girls' cultures and femininities in the primary classroom. *Gender and Education, 13*(2), 153–166.

Robb, M. (2007). Wellbeing. In M. J. Kehily (Ed.), *Understanding youth: Perspectives, identities and practices* (pp. 181–214). London: Sage/Open University Press.

Schooch, R. (2007). *The secrets of happiness: Three thousand years of searching for the good life*. London: Profile Books.

Sennett, R. (1998). *The corrosion of character: The personal consequences of work in the new capitalism*. London: WW Norton.

Sennett, R., & Cobb, J. (1977). *The hidden injuries of class*. Cambridge: Cambridge University Press.

Shildrick, T., MacDonald, R., Webster, C., & Garthwaite, K. (2012). *'Poverty and insecurity' life in low pay, no pay Britain*. Bristol: Policy Press.

Sustrans. (2012). *Locked out: Transport poverty in England.* Sustrans. Retrieved from http://www.sustrans.org.uk/lockedout

Sutton Trust. (2014a). *Analysis of trends in higher education applications, admissions, and enrolments.* London: Sutton Trust.

Sutton Trust. (2014b). *Summary: Attainment gaps between the most deprived and advantaged schools.* London: Sutton Trust.

Sweetman, P. (1999). Anchoring the postmodern self? Body modification, fashion and identity. *Body and Society, 5*(2), 51–76.

UNICEF. (2007). *Child poverty in perspective: An overview of child wellbeing in rich countries.* Florence: UNICEF.

Veenhoven, R. (1999). Quality of life in individualistic societies: A comparisons of 43 nations in the early 1990s. *Social Indicators Research, 48*, 157–186.

Wasserman, D., Cheng, Q., & Jiang, G. (2005). Global suicide rates among young people aged 15–19. *World Psychiatry, 4*(2), 114–120.

Webster, C. (2009). Young people, 'race' and ethnicity'. In A. Furlong (Ed.), *Handbook of youth and young adulthood: New perspectives and agendas* (pp. 66–73). London: Routledge.

Webster, C., Simpson, D., MacDonald, R., Abbas, A., Cieslik, M., Shildrick, T., et al. (2004). *Poor transitions: Social exclusion and young adults.* Bristol: Policy Press.

WHO. (2009). *Child and adolescent health and development 2008.* Geneva: World Health Organization.

Wilkinson, R., & Pickett, K. (2009). *The spirit level: Why equality is better for everyone.* London: Penguin.

Willetts, D. (2011). *The pinch: How the baby boomers took their children's future—and why they should give it back.* London: Atlantic Books.

8

The 'Thirty Somethings': Happiness in the Late Twenties and Thirties

In the previous chapter we explored some of the ways that happiness is experienced by young people and how structural constraints such as the availability of jobs and housing can be a barrier to good wellbeing. The sorts of socioeconomic resources that young people have access to usually mediated through family relationships can be influential in framing the happiness of people as they grow up. There is not, however, a simple relationship between socioeconomic background and wellbeing as our younger interviewees showed, some have extensive social networks and participate in cultural activities that can bolster their wellbeing, despite the incidence of material disadvantages. This complex nature of wellbeing was also evident amongst the older people I interviewed for this project—as we now see with the 'Thirty Somethings'.[1] There were examples of the significant role that economic resources or social class had on people's happiness yet interviewees' wellbeing was influenced by the shifting dynamics of family/friendship relationships too.

[1] Although I refer to his group as the 'Thirty Somethings', two of the sample were in their late 20s, but much of the talk at interview, even with these younger interviewees centred on becoming a 'Thirty Something', hence the title of this chapter.

M. Cieslik, *The Happiness Riddle and the Quest for a Good Life*,
DOI 10.1057/978-1-137-31882-4_8

The abilities of respondents to manage or cope with the challenges of life also created complex patterns of wellbeing that worked against any simple generalisations about the role of class or gender in the structuring of happiness.

The Thirty Somethings sample comprised five individuals. Sally was 31 when first interviewed in 2010; her parents were both university educated, working in professional occupations, and she had graduated with a degree in anthropology from an elite university. Jenny was also 31 when interviewed in 2011; from a working-class background, her parents employed in semi-skilled occupations, and she had left school at 16 years of age with some passes at GCSEs. Barry was from a middle-class background as his parents were university educated and had worked in professional occupations, though he had dropped out of his university programme before graduation. He was 27 at time of interview in 2011. Carl who was 28 years of age was from a working-class background, his mother working in a clerical occupation and he had worked since he was 18 after completing some vocational training. The final sample member was Katherine who was 31 at interview in 2014. Her parents worked in semi-skilled occupations and she had left school with some GCSE qualifications and worked in unskilled jobs. She took a university access course in 2012 and was studying at university at the time of interview.

Although there were similarities in how the younger sample and this older group experienced happiness there were also marked differences. In particular, we can now see in the older group some of the consequences of those earlier experiences they had when teenagers. This is one advantage of a biographical approach (Hockey and James 2003), in its ability to trace some off the cumulative effects of life events on wellbeing—a feature often neglected in happiness studies with its use of snapshot surveys. This approach allows for different aspects of wellbeing, such as anxiety or more positive orientations, to be linked to earlier life events. This older group has also had greater time to reflect on earlier experiences, and this also has influenced their ideas about happiness and the good life when compared to the younger participants. This does raise some interesting questions about whether we become wiser with age and experience that offers us greater insights into how to live well.

Friends and the Meaning of Happiness?

A common theme across all age groups was that interviewees spoke about happiness in the popular, common sense way of meaning simple, everyday experiences that evoked positive emotions. Spending time with a partner, eating a meal, a visit to the cinema or glass of wine all featured in accounts of popular activities that were seen as important to having a good day. 'Being happy' as Carl discussed was understood as being able to maximise these sort of good, nourishing events in life that produced positive, subjective feelings.

> Leisure is important, we have a night out every week, some [friends] are from school and others are from bands I have been in ... we always end the night talking about music and local bands ... we all grew up on estates so we all had the same experiences of growing up. (Carl)

Interviewees acknowledged during our conversations that many of these positive experiences were dependent on good relationships with others—their partners, friends and work colleagues. Behind the apparently simple acts of having some fun were more complex webs of relationships, the quality of which played an important role in the patterning of these positive experiences. This is a feature noted by earlier sociologists such as Durkheim in his classic studies on Suicide (1984) and the Division of Labour (2014) where relationships offer the regulation and integration, essential for good wellbeing. These friendship networks were ones that had taken many years to establish and people often overlooked this historic backdrop of their current experiences. People have to work at these friendships, making an effort to keep in touch and participate in activities and reciprocate the support they offer. Behind the façade of a night out and what seem simple events that support wellbeing is a long history of practical effort and activities that sustain these networks.

When pushed to describe what it was about friends that were important to wellbeing, interviewees spoke of 'being able to be themselves', 'to relax', 'sharing interests, views or experiences'. Erving Goffman (1956: 66–69) suggests that time with friends is valuable and nourishing as these are safe spaces or 'back stage' arenas where one has some respite from the public roles and performances that we undertake at work, school or the street. We are not expected to monitor our speech, dress and thoughts quite as much with friends as we are in these other 'front stage' spaces. Research by sociologists such as Ray Pahl (1995, 1998) points to the support that friends offer, functioning as a resource that can help people cope with the daily challenges of work, partners and children. These simple things of having someone to talk to, encourage us or to offer a hug help us to understand how friendships function to support wellbeing. Jenny spoke of the importance of friends for her happiness.

> I have three best friends, two of them have kids, Ruby doesn't have kids she lives at home with her mum and dad who are deaf, she's my bestest friend … I go around there and just chill out, put telly on and we have a few vodka and cokes and she sticks pizza in the oven and we just talk shit all night about stuff and I roll in at midnight a bit drunk and smelling of fags … Ruby and I are both headstrong and she'll say, 'Jenny you know I love you but you are such a cock sometimes', you know like if she says something I don't agree with, I'll tell her and she does it to me but that's part of being best friends isn't it. (Jenny)

When asked what she talked about with her friends Jenny remarked, 'that it was just everyday stuff'. Friends then are such a familiar part of everyday routines that they are often difficult to analyse to see how they underpin wellbeing. This is a challenge for those trying to make sense of wellbeing as so much of it is rooted in innocuous, taken for granted routines (Dolan 2015; Kahneman 2011; Mills 1959). Respondents' accounts of nights out with friends were characterised by mundane chat—talking and laughing about some of the hassles of everyday life—disagreements with partners, problems with the behaviour of children, worries over money, problems with neighbours and so on. This time with friends was valued as it was a chance to offload to others who are trusted and supportive, sharing one's stories of trying to get through the week. Friends are important as they offer advice on decisions that have to be made—getting a second opinion from someone is significant for the choices we make about how to live well—the everyday ethics that frame our lives (Sayer 2011). To be helped to make informed choices and have support from people that we know and trust can help us make better choices that hopefully enhance rather than undermine our wellbeing. But a few hours in the pub or around a friend's house was also an important change to the usual routine of work, partners, children and running the home. Over two millennia ago, Aristotle wrote of the significance of friends to a good life and he reminds us of the need that we all have to be sociable and to be valued by others (Aristotle 2009). It is through friendships that we often see altruism and compassion flowing between people and helping them to a balanced wellbeing—which is a form of social glue, helping us to bond with others (Gilbert 2005). Many commentators in recent years have documented the role of social capital to healthy communities and happier individuals (Putnam 2001) and we can see here in relation to friends the importance of apparently simple relationships to people's wellbeing. One can see how social isolation where one lacks these sort of regular contacts with others or social capital can be so corrosive of wellbeing and hence why policy-makers and practitioners promote wider social networks to improve the quality of life of people today (Age UK 2015).

Happiness, Solitude and 'Time for Oneself'

Despite the reliance on social networks for positive experiences, our sample of 'Thirty Somethings' also spoke of their involvement in quite solitary activities that they felt helped bolster their wellbeing. This illustrates another riddle of happiness—how it is multi-facetted so that many people enjoyed spending time alone whilst also depending greatly on good social relationships for 'good feeling' and contentment with life. Interviewees were involved in solitary pursuits as they enjoyed them and they had become part of their routines, rather than them emerging from a self-conscious quest for better wellbeing. Barry, for example, discussed his interest in gaming—using game consoles most days, playing solo or online against other people on the web. He spoke of the enjoyment of mastering the different levels in these games and how he could, 'lose himself for hours at a time'.

> I have an Xbox and Play Station and I have been brought up with them. And I suppose it is one of the things that keeps me sane, at certain times …

> It gives you a chance to shut your brain down, when life is really, really crap … It is almost like pressing a reset button on your brain sometimes. It is removing yourself from it, if you play the right kind of games, things that take over all your senses, shut off what is actually going on in your own life for a while and you give yourself a break from it … I think you can use it as a barometer of my mood, how much I play on them. The worse life is the more time I spend gaming. (Barry)

Gaming had become a part of Barry's daily routine, spending a few hours sat at his computer playing one of several games which he said helped him to unwind after a day at work, driving his taxi. It was a form of escapism but was also therapeutic (Csikszentmihalyi 2002) as it was not a passive activity like watching TV for Barry spoke of having to engage his skills, thinking through strategies and being challenged when playing these games—all of which are important to the enjoyment or 'flow' he derives from gaming. Sally in contrast had developed a more traditional interest in drawing and painting as a way of relaxing, depicting the landscapes she encountered when walking near her home. These images of rivers, valleys and mountains were dotted around her house—reminders of those tranquil and enjoyable moments she had over the years. These pursuits she said were rewarding as she had developed her skills over time and often it was challenging trying to capture some of the scenes she encountered on her walks. Another respondent, Katherine, had developed an interest in photography, taking pictures of the different places she had been walking and she brought in many of these images to the interviews to illustrate what it was she enjoyed about walking and how it contributed to a positive wellbeing.

> This is the picture of the park where I go to walk and these are the types of views we get. I'm just sat there … You can see here a picture of the Northern Lights … Happiness is being clear in my head and being with nature clears my head. It clears everything. (Katherine)

The images documented the many visits to her local nature reserve over several years. As the site was elevated she was able to make her way through the woodland, every so often glimpsing through the trees the patchwork of fields and hedgerows for miles around. The images showed how the crops changed from one year to another and how the trees and flowers evolved through the seasons. There was something about the quiet, peaceful setting of the park and the beauty of the trees, flowers and birdsong that Katherine found restful and restorative. For many years Katherine had struggled to make ends meet, surviving on social security and often worried about providing a good life for her children. Many of these negative thoughts that she experienced had come to frame her life in recent years and so the walks she took offered a temporary respite from some of these anxieties—to clear her head. This is an understanding of nature and its positive effects on wellbeing that many philosophers have noted from Aristotle (2009) to Marx (1983) and Nussbaum and Sen (1993). Although we often regard ourselves as separate from nature, we in fact have an elemental link to the natural world that we need to foster to live well. Barry too, despite his interest in gaming, also talked of nature as restorative and nourishing and as a form of escape—this time from some of the challenging experiences he had working as a taxi driver.

> If you go up the main road the countryside opens up and you can see the hills in the distance—it's brilliant for just going, 'Ah [sighs], I'm just so tiny again'. Sometimes I go to the seafront. A few weeks back, there were a couple in the backseat, really racist, loud-mouthed. I tried to close it all off, got irritated by it so dropped them off, got my money and pulled into the Dene and you can look down over the town and all that big space. I had a cigarette and a drink and just watched the world go by for a bit. Watched the seagulls playing with the wind and was happy again. Got back in my cab and went back into town for my next fare, I suppose I use it for a kind of medicine. (Barry)

Nature or experiencing the countryside in these examples functions as a therapeutic break from a busy life, offering some balance or counterpoint to the frenetic pace of work and family life. The trees, grasses, flowers and wildlife were also seen as having an inherent beauty that generated a sense of awe and wonder about the natural world that was nourishing. These were experiences that older interviewees also discussed when talking about the beauty in everyday life—those odd sublime moments that help one to have a good life (Schiller 2004). Some interviewees took this beauty at face value whilst others pondered this, further speaking of how as humans just like other animals we are connected to nature, all sharing the Earth yet we often forget these links. For Barry a regular connection with the countryside was important as he felt we needed to counter the effects of busy lives that alienate people from these elemental connections to nature—we run the risk of losing a grounded sense of ourselves and get caught up in needless consumerism and yearnings if we neglect our relationships to the natural environment.

These solitary pursuits—gaming, drawing, photography and walking—functioned, like the more social activities we have discussed, to support wellbeing as they offered an escape from the usual routines of everyday life. A good life in part is made up of regular opportunities to free oneself from the daily grind of waged work, home and caring responsibilities. The chance to lose oneself in an imaginary world created by computers or a beautiful world of nature fulfils an important need we all have for some sort of escape from ordinary routines and expectations that others have of us. If these escape activities also allow us to be challenged, to learn and develop skills then as we see with photography, gaming and painting as Csikszentmihalyi (2002) documents these pursuits can be particularly nourishing for our wellbeing. Unlike those more sedentary, passive activities like watching TV, 'flow' experiences meet a fundamental need in all of us to learn, to create and be challenged and this is why so many interviewees spoke so positively about their hobbies and interests.

Aging, Independence and Wellbeing

One significant difference between the Thirty Somethings and the younger cohort of interviewees was that the older group had established long-term intimate partnerships and were sharing a home and this offered emotional stability and a sense of independence from parents. They all spoke about growing up and being more comfortable with who they are as they approached their mid-20s and moved towards their thirties. This illustrates an important dimension of happiness and living well—namely the desire for self-discovery and self-development (Taylor 1991). Taylor develops arguments seen in Marx (in his work on non-alienation) to suggest that as humans we have a calling to seek out new experiences and personal growth. We see this with the Thirty Somethings as they seek out partners, homes and parenting for the first time. At the heart of these transitions and new experiences is the desire to love and be loved—a feature of wellbeing that, though recognised, has been under-researched (Hyman 2014; Smart 2007; Thin 2012). These transitions involve important choices about how to live and force people to reflect on their values, goals and expectations. Such transitions from lives as singletons to couples, however, were not always easy—all spoke of failed relationships and heartache but that such disappointments, though at the time were difficult to endure, were part of growing up. This contrasts markedly with some sociological accounts (Marcuse 2002; Furedi 2004; Davies 2015) that suggest that happiness is often experienced as simple positive emotions. Instead, from our interviewees we have a more nuanced understanding that a good life is not about avoiding loss and suffering but entails an acceptance that life is inevitably painful and one has to deal with it as best as one can. My interviewees suggested that out of these hardships comes an appreciation of the good things in life and a more rounded and experienced understanding of how to live well—a more sophisticated self and a richer life.

The significance of the intimate relationship for wellbeing did vary nevertheless between the members of the Thirty Somethings group as different individuals were at various stages of this life course transition. Carl, Barry and Sally had only recently set up home (within the last year) with their long-term partners and so the fun and novelty of this meant they often spoke of being very happy about their domestic lives and relationships with partners. By contrast Jenny and Katherine had been cohabiting for several years and admitted that the novelty of 'domestic bliss' had worn off. This experience of sharing a home and its positive contribution to wellbeing has been noted by research (Layard 2005: 65–66). Interestingly, despite such accounts both Jenny and Katherine offered a more nuanced perspective, as they felt more positive about their current home life and wellbeing because of difficult past events they had overcome. Katherine spoke about the breakdown of a previous long-term relationship with a man who had helped raise her two children but that out of this difficult period she had come to terms with her bisexuality and forged a new relationship with her current partner, Karen.

> My ex was quite controlling, 'why are you wearing perfume'? 'You going to meet someone'? Just little things … but I found out why he didn't trust me as he was sleeping with my best friend for seven years and they had had a kid together … At the time I was size zero, stopped eating, cut my hair off. It was the worst time of my life. I thought I was worthless. You know you get shit on from afar, it was absolutely horrible, but there again it was an experience which has helped me … I know it sounds really stupid but I took a camping trip to the forest to make me realise, 'What am I doing'? It was me, the girls [two daughters] a few friends and Karen. If it wasn't for Karen I don't know if I would be here … Sometimes I think, 'What the frig do you see in me'? 'What did you see in me then, especially then'? (Katherine)

The breakdown of her relationship was a low point in her wellbeing but that she was able to rely on her friendship network and her love for the countryside to begin the process of moving on. In particular, finding a new partner Karen, an old friend from school, was a pivotal moment in helping to establish a new home and family with her children. Although satisfying, waged work and good friends and neighbours featured in respondent's accounts of a good life they all also spoke about the centrality of a good, loving relationship with a partner for their happiness (Thin 2011). Finding someone to love and to be loved, someone to share one's life with was at the heart of many people's stories of their lives. And it is when these relationships break down that interviewees spoke of the sadness, regrets and heartache that then featured in accounts of the struggle and suffering that life often brings and which we have to manage as best we can. These stories illustrate the inherently social and emotional nature of happiness and a good life and how we need to use narratives of these experiences to make sense of them—as much as the numerical expressions employed by the countless surveys of wellbeing.

Jenny's story shares some of the features of Katherine's biography—notably the difficulties posed by a controlling male partner and how one can overcome these difficulties to be happier. We see here some of the complexities of what a happy life means to young women like Jenny. Like many working-class women she grew up hoping that she would find a partner with which to share her life—to have children, a home and career. But the reality of her life through her twenties was rather different.

This is an example of how some sociologists such as Sara Ahmed (2010) construct happiness—as sets of norms about how to live that are often unobtainable and lead to disappointments and sadness—the 'problem of happiness'.

> Phillip [partner] and I split as he wasn't giving me any money. He was paying the bills but only little bits of it to me. I was sick of it—just being a mum—no independence and no money. Shit existence, fed up asking him for money all the time. If the kids were sick and I couldn't do my job he didn't make the money up … He thought I should be just grateful with a nice home and the bills paid. But I didn't want to spend the rest of my life with someone who was greedy. He just didn't understand the cost of the kids. After two years of that I thought stuff it... so I said to him, 'Off you go'. I didn't plan to split it just came to a head… The last few months he has tried to make it up so I have said to him that I need housekeeping every week, a direct debit and that I want to go back to college and he has to support me or I'll bugger off. He's been fine about it so far, I'm going to wait and see if it's alright. (Jenny)

Jenny did eventually allow Phillip to move back into the family home under these new conditions. She hoped he would be more understanding of how difficult it was to run their home. She was hoping that the marriage would be more of a partnership and this would make for a happier life for her, the children and her husband. We see in Jenny's account a mismatch between her aspirations for a good life and reality of the opportunities she has. Being happy for Jenny involves a struggle between the differing ideas that she and her husband has about how to live. During this time Jenny has tried to negotiate with her partner over the different elements of their lives—finances, domestic labour, parenting and so on. We see how happiness is woven into the fabric of everyday struggles, trade-offs and balances of power between people.

We also see in these examples the riddle of happiness as a good life depends on having experienced loss, suffering and disappointments so that one can understand what a better way of life actually is. Having had difficult times in the past often involves slowly working through a process of finding a different way to live—developing 'practical reasoning' (Aristotle 2009; Sayer 2011) and taking small steps towards a better life. Past events that had been problematic can help one to make more informed judgements and help us in efforts to live a more contented life. For Jenny, she did precisely this, around the way that money and housework was shared out in their relationship so that it was more equal. Yet, curiously, much wellbeing research neglects this way that ordinary people navigate their way through these challenges, trying to overcome problems as they try and lead happier lives. In fact, much writing on wellbeing suggests that people view suffering and disappointment as pathological and to be avoided, seeing a good life as one that only comes from a focus on positive experiences (Ahmed 2010; Furedi 2004; Cederstrom and Spicer 2015; Davies 2015). Yet accounts from my interviewees would suggest otherwise—all understood that suffering was a part of living, even if one wished to avoid it, in time it could enrich one's life.

Happiness and Work-Life Balance

One argument in this book is how happiness is socially situated—that happiness and its sources varies greatly, depending on the characteristics of individuals and the social settings in which they live. Thus the same experience, such as a day at work or a night out with friends may have very different consequences for the wellbeing of two different people. Because of their biographies and different perspectives they bring to a situation, different individuals can experience happiness and its sources in distinctive ways. For someone disenchanted with work, another day at the office fills them with dread but for someone entering work after years of unemployment it can be a joyous experience. For many young people who have much free time, a trip to the pub is an innocuous, everyday affair but for older individuals, working long hours with family commitments, it is something to be savoured. We see these differences when we compare the views of the young people sample with the older cohort of interviewees in their thirties. There is something about aging that confers poignancy to what were taken for granted experiences because of the increasing demands placed on one's time. Katherine, who was just finishing her first year of a degree, found it very difficult juggling the demands of exam revision with those of running her home and seeing her children and her partner. She writes at length in her diary about how April and May had been very hectic and stressful, in part because she had assessments to submit yet these deadlines coincided with the school holidays so was caught between looking after her children and the need to study.

> Sat in my bedroom. Feels like I've been stuck in my bedroom for days and it's only been one day of exam prep. It is quite funny really when I think about it I can feel the happiness drain from me. It's not the work I'm doing that is draining as I love learning but the constant drabness of these four walls, of being stuck in the house continuously.

> Today was a fantastic day—'the best things in life are free'. Went to the beach with a picnic and a football and had great family time … It was such a great afternoon and made me realise don't forget to enjoy the little things in life. (Katherine, diary entry)

Where previously time on a games console or night out with friends would have been easy to manage, now such opportunities have become increasingly difficult to organise—longer hours at work, the demands of young children and the wishes of partners had all encroached on interviewee's spare time. Many of this older group were nostalgic about their free time when young and felt they had taken this for granted, unaware of how busy they would be when older.

For the two women I interviewed who had young children, they held strong views about the challenges of juggling their many different responsibilities and its impact on their wellbeing. Jenny had recently started an education programme at college and she was juggling the demands of this with part-time work as a cleaner and the needs of her two young children. In the midst of all these things she felt anxious and stressed about how little time she had for herself, partner or friends—yet these were important for a good life and happiness.

> I just get stressed juggling things. Some people enjoy it don't they, the pressure of things and juggling things but it just makes me quite stressed … Life is just one big slog isn't it? It is, because I find life quite hard really… Life's a struggle all the time … I never have time to myself … that never

> happens, life is just busy isn't it. I mean I am in bed at 9pm because I am just exhausted with the day. (Jenny)

Recent decades have seen research (Shaw et al. 2003; Crompton and Lyonette 2006; Lunau et al. 2014) that suggests that people feel increasingly stressed and pressured because growing demands on people's time is eroding the work-life balance. However, in Jenny's case it was not long hours in employment that undermined her wellbeing rather that overall there were too many competing demands on her time, particularly as a 'caregiver' (for her children and partner). She felt that she was juggling lots of different activities each day. Her views on the lack of balance in her life reflects various research that points to the 'time squeeze' or 'time crunch' experienced by many women who try and balance the needs of children and partners with demands of waged work and domestic responsibilities (Roberts 2007: 339).

Women in particular have been shown to suffer from expanding responsibilities (such as care-giving) and poor work-life balance and in turn poorer wellbeing when compared to men (Crompton and Lyonette 2006). This gendering of poor wellbeing and work-life balance can be attributed partly to the unrealistic expectations society has of women to combine roles as carers with those in the labour market. Whilst by contrast men often shrug off these domestic and caring responsibilities. This social structuring of poor work-life balance for women in the UK has also been exacerbated by cuts to State provision of social care (Guardian 2014). The women I interviewed felt responsible for the wellbeing of their partners and children—they saw themselves as working for the happiness of others much of the time—a notion that was seldom shared by the male interviewees. The mothers in the sample spoke at length about being worried about their children's safety and happiness. They also spent much time concerned about the welfare of partners and elderly parents. This caring meant numerous school visits, children's parties, help with schoolwork as well as time and effort maintaining relationships with their own parents. If grandparents were able to they could offer support themselves with childcare but with time female interviewees spoke of their increasing role supporting their own parents to ensure they had a good quality of life. These accounts reflect the work of sociologists such as Arlie

Hochschild (2003) that point to the many different ways that women's lives involve emotional labour, nurturing others to ensure their wellbeing yet this is often hidden from view obscured by how such activities are unpaid and appear 'natural'.

This gendered division of labour around relationships, emotions and responsibility for wellbeing does not necessarily have to create problems of work-life balance for women. In many Scandinavian societies women score more highly in wellbeing surveys than in the UK and researchers point to the state provision of paid parental leave and subsidised childcare as key to these differences in wellbeing (OECD 2002; Esping-Anderson 2009). Such welfare services allow women (as well as men) to have far more time to care for children and then to combine work and caring which prevents some of the work-life balance issues we see in the UK. In contrast, women in the UK, as we see with Jenny and Katherine, received relatively small amounts of State support whilst on maternity leave and were thus compelled to return to work. But then they were faced with expensive childcare bills that consumed much of the income they received from the poorly paid jobs they undertook. These are some of the important structural constraints that frame the lives of many women in the UK today and which have emerged over many years of policy reform driven by a system that is market led rather than the social democratic approach of Scandinavian countries. This is one way that shifts in policy over many years can come to influence the wellbeing of individuals in different societies and the example of Nordic countries poses important questions about how best to change policy in the UK to promote better wellbeing for women and their families.

Social Class, Austerity and the Structuring of Wellbeing?

Since the turn of the century there has been a wealth of research that points to the widening gulf between different sections of society in the UK, particularly how the declining quality of employment coupled to the erosion of welfare support (such as unemployment benefits, health, social care and education) has fuelled social inequality and impoverished many (Wilkinson and Pickett 2009; Clarke 2014; Toynbee and Walker 2015). Any study into wellbeing then has to examine how these processes of marginalisation have impacted on the quality of life of people in the UK. For our small group of 'Thirty Somethings', there were stark differences in the life course routes that the working class took (Katherine, Jenny and Carl) to those from more affluent backgrounds (Sally and Barry). For sociologists, these differences in transitions experiences and opportunities are key to understanding many of the differences in the quality of individual's lives. However, although material conditions and welfare systems are significant, interviewees also discussed other experiences discussed other experiencesthat influenced their wellbeing which illustrates one of the key arguments of this book—namely that happiness is multifaceted and has multiple sources.

To demonstrate the complexity of happiness we only have to compare the working-class Jenny with the more middle-class Sally. Here we can see that differences in parental resources do profoundly shape the routes that people can take through life and how such differences may influence wellbeing. Jenny grew up on a disadvantaged estate and many of her friends were disinterested in education so she became part of an anti-school girl gang who truanted. Jenny's parents were happy for her to leave school and find work in local shops or nursing homes—replicating the sort of experiences they themselves had of education and work. The cultural assumptions underlying some working-class families is that some modest achievement at school and then transition into work and a wage will allow for relatively happy working-class life and this is how Jenny spoke of her early years of adulthood. However, by the age of 23 Jenny had established a serious relationship, becoming pregnant shortly afterwards and then moving into social housing before the baby was born. Though Jenny spoke of the joy of becoming a mother and setting up home, these early years of cohabiting and motherhood were financially challenging as Phillip her partner was in a low paid manual job and she earned very little in her part-time job. Jenny spoke of how her wellbeing at this time was rather mixed—having a partner, child and home signalled the transition to independent living—a positive development compared to the life at home she had as a young teenager. Yet she also spoke of the stress and anxiety caused by running the home with little money and relying on her partner's wage to meet most of the bills. Though it appeared she had a better life she felt she had merely swapped one form of dependency, on her parents for another—dependency on her partner. As she got older Jenny discussed how she felt unfulfilled in her life and wanted something more than being a mother and partner.

> I've been cleaning toilets for a living and I don't want to do this anymore—it's getting me down. I don't want to clean for people as I have more skills than that ... I've started this access course; just felt I needed to do something for me. I was fed up being just a mum ... it's about fulfilling my career because I think I am wasted really. Not blowing my own trumpet but I think that one of the happiest times in my life was when I worked on a busy ward and I think I did very well on it. (Jenny)

The wellbeing questionnaire that Jenny completed reflected many of the issues she discussed in the interview. She rated her satisfaction with work, home and main relationship at only 4–5 (on an 11-point scale) and that she was unhappy 65 % of the time and happy 25 % of the time—when asked how satisfied she was overall with her life she rated it at 4 out of 10. Many social scientists would understand Jenny's predicament by suggesting that her bouts of poor wellbeing stem from the poor resources and opportunities (such as low income) of women growing up in working-class families (Ackerman and Paulucci 1983; Aknin et al. 2009). Because of her poor start in life the jobs she took were poorly paid and unfulfilling which then shaped the quality of her housing and community and in turn her wellbeing. At each key point in her biography such as access to schools, examinations, the search for employment and housing she was in competition with others (Bourdieu's notion of field) and her material and cultural resources influenced her ability to make transitions to adulthood. Of course what we are unable to witness by only interviewing Jenny are the other people such as classmates, neighbours and job seekers who enjoyed better resources and success in the competition for credentials, housing and employment.[2] Researchers suggest (Jahoda 1982; Tait et al. 1989) that the more affluent enjoy greater parental resources and

[2] One weakness with small-scale qualitative projects such as this is the inability to adequately map the social networks of individuals (and their structure or fields) to explore in detail how actors

these help individuals to educational success and employment that is fulfilling which frame better wellbeing. Certainly, when I first interviewed the more middle-class Sally she spoke of being fortunate as her parents, who were both university educated, were very supportive of her education, ensuring she attended a good school and helping with her studies and ambitions for university and eventually a career of her own. There is something about the cultural environment of some middle-class homes, the ambitions and expectations of parents that can work positively for children as they grow up, aiding their transition through school and into adulthood—features which some working-class families may not always have. Sally then was encouraged to be independent and to question the restrictive, gendered assumptions about what women can do with their lives.

> When I was younger when I was growing up I was allowed to do what I wanted and there was always room to negotiate with them [parents] and talk about things … They brought us up to be really free thinking and free people … On my gap year in Africa, living in that mud hut in the middle of nowhere I became really resilient … When I was away in Tanzania, they had nothing and they were the happiest people I have ever lived with and I learnt a lot from them. (Sally)

The contrasts between these different women and their wellbeing reflect the social divisions in Britain. To account for differences in wellbeing through structural differences is a traditional way of understanding wellbeing for many researchers (Wilkinson and Pickett 2009)—focusing on the links between particularly material resources, cultural upbringing or habitus (Bourdieu and Passeron 1990), education and labour market transitions and quality of life. Yet at interview the women suggested there were also other features of their lives that had a powerful effect on their wellbeing. Notably, Jenny spoke of how despite the material difficulties she has endured, she also felt that she had grown up markedly with her transition to a homemaker and becoming a mother. She felt a sense of achievement and status in raising happy, contented children.

---

deploy various resources as they pursue their goals. To do this one would need to interview many more participants which was beyond the scope of the current project.

> I've got quite a good life you know. I have made some achievements. I have my own home. I know it's in the middle of a council estate but it's ours you know ... We eat well and the children have the best of things ... And Phillip [partner] loves the kids, he's a good dad, we want the same thing for the children. (Jenny)

Although on the surface it may be seem that Jenny's poorer material situation and difficulties in waged work would determine her wellbeing, she suggested otherwise. Despite feeling unfulfilled in terms of a career, she also felt a real pride in the 'simple' achievements of working at her relationship with Phillip, raising her children and running the home.

Follow-up interviews with Sally also produced a more complex picture of her wellbeing. Though undoubtedly her educational success and career were sources of satisfaction—framing her identity as an independent women, in recent years she has also become conscious of other features of a good life that had so far eluded her. For some years she had witnessed (through online social media sites such as Face Book) many of her friends settle down, marry, and have children and she was unsure about this next step in her life—as was her boyfriend.

> It is getting to the point in my life where I have to start thinking about what my future is ... lots of people are marrying off and settling down ... Crunch time. I have got my family saying, 'don't bother with him', him going, 'I am too busy', and I have got me going, 'probably should make a decision soon'. I am a little lost ... Well, I have got a strategy, yeah. Stay with boyfriend 'til September ... and then after then, I mean if he doesn't want to commit by then, then I am going away. That's true! Otherwise I'm having marriage, I'll have that, babies or kids ... think he knows that I'm kind of coming to the end of my tether with it. ... I've talked to my mum and dad about it ... they thought I was the most unhappy I've ever been? (Sally)

Paradoxically, Sally was envious of our working-class women and the 'simple' things they had achieved. Finding a long-term partner, setting up home and starting a family certainly did not seem like a simple challenge but she felt it was now the key to a good life. Once again we get a sense of the way that societal norms about happiness are internalised as we grow and at times come to frame our wellbeing often in corrosive ways (Ahmed 2010). At the same time, Jenny and some of the other working-class women I interviewed were envious of middle-class women like Sally—they too would like a university education, a better paid, more enjoyable job and chance to live an independent life. This is another example of the paradoxes of happiness—how other people's lives can seem more appealing than our own—'how the grass always seem greener on the other side of the fence'. Our judgements about our wellbeing often involves comparisons with reference groups (or social comparisons) and accounts for why many are restless or dissatisfied with life as we continually upgrade our expectations of a good life (Layard 2005: 44; Dolan 2015: 139–140).

Both sets of women were looking wistfully across the class divide that exists in Britain today, feeling disappointed at what was missing in their lives and dreaming of different futures. Although the material features of people's lives undoubtedly frame their wellbeing, at the same time how people judge the quality of their lives reflects the wider expectations they have of life as they age. This not only includes expectations about jobs, income and lifestyles but also about those other key features of life—finding a partner and being loved, raising a family and creating a home. Sally had on the whole been happy throughout her twenties as one might expect but then increasingly become dissatisfied with her life as she moved towards a period when she expected to settle down and become a mother. A better future for Sally was one that featured the opportunity to have what the working-class women had already experienced—children and a home with a partner. For Jenny her twenties had often been a struggle yet she felt she had achieved something important through motherhood and home making. But then she also felt that these years had been marked by their dependency on others and longed for the opportunity to develop her skills and a more secure future. For Jenny a better future was one that offered some genuine independence just as Sally had had—a return to education, university and a middle-class career and income. In many ways these women were experiencing an early or mid-life crisis where aging and shifts in their circumstances had prompted them to re-evaluate their lives and how they looked back at their biographies and forward to their possible futures. The ways that individuals reflect and then reframe aspirations and expectations does pose difficulties for how sociologists undertake happiness research. For these life crises show how tenuous sometimes the connection is between objective circumstances of individuals and their subjective evaluations of wellbeing. Where jobs, partners and homes at one point may lead to very positive evaluation of wellbeing, a few years later individuals may have developed far more critical and negative judgements on their lives.

Making Sense of Key Moments in Life: The Conflict Between Habitus and Personal Change

What was interesting about the Thirty Something women, irrespective of their material circumstances was how they had experienced periods when they felt 'stuck' or conflicted about their lives and how best to make choices about the future. All had spoken of having long periods of indecision and low-level anxiety about how best to change their lives and flourish. One way of understanding these difficulties is to view them as emerging from a conflict between the established ways that people have of thinking about themselves (that reflect their upbringing or habitus) and the new challenges that individuals face as they age and grow. My interviewees had grown up with distinctive assumptions about themselves—the sorts of jobs, families and partners they would have—but with time some of these have become outdated as they are presented with new opportunities and experiences. As many other writers have discussed (Williams and Penman 2011; Deary 2014; Dolan 2015) for individuals to flourish they have to monitor and change the assumptions and routines that underpin their lives, otherwise these habits can become a barrier to a good life. But to challenge and transform deeply held ideas can often be very difficult and unsettling.

Sally, who was a middle-class woman, had grown up with values that encouraged her to live an independent life and to cherish success in education and work. From an early age she said that she wished to enjoy life as much as men—playing sports, spending time walking and exploring the countryside—as a result she was often seen as 'tomboy', as she challenged the accepted ideas about femininity. However, the difficulties she encountered in recent years were how to make the transition from the life of a single woman to one that featured a long-term partner, home and children. The difficulty was that this entailed a change to some of these ways of thinking about herself and how she lived. She spoke to me of being unsure about what would be best for her happiness. Though she loved her boyfriend, he too seemed unsure about whether marriage, a house and children together would work. Sally was also concerned that she might lose some of her independence—she liked her career and her salary. If they were to have children could she give this up and rely on her boyfriend to support her? In the past when faced with difficult choices Sally usually consulted her family about what to do. However, her parents were unsure about her boyfriend and they also worried that she would lose her independence. They also failed to appreciate how much Sally loved her boyfriend so their advice she felt was often unhelpful. She didn't seem to have the skills or experience to make an informed decision about what to do. These issues came at a difficult time as her contract at work was up for renewal (she had been working on a three-year Arts project at a museum) and had some health problems that had added to her sense of distress (she had been diagnosed with anaemia).

In contrast, Katherine's dilemmas about her life and being happy were very different in substance from Sally's but were similar in originating in tensions between traditional assumptions and the new challenges posed by aging. When interviewed Katherine was at the end of her first year at university studying criminology. She was enjoying the course but felt that she did not fit in with the other students for she was a mature student from a working-class background unlike the many teenagers and middle-class students on her programme. For Katherine, the past year had been challenging as she had been disappointed about the impact that attending university had had on her life. She had returned to education as her children were now older and she felt that a degree might help her to use some of her skills and develop her talents. She had always been interested in politics and current affairs and although she enjoyed being a mother and running the home she also felt that there might be more to life than being a mother.

Over the course of several interviews we were able to unpack some of the reasons why Katherine felt so conflicted now about her current situation—why she was disappointed about being at university when it had seemed like a positive step for her in life. She had struggled in school when young suffering with dyslexia and so had moved into various unskilled jobs and then parenthood as a teenager. She grew up with narrow, restricted ideas about her talents and what her life may bring—her habitus involved a lack of confidence about her academic abilities and modest ambitions about jobs and careers—a feature widely documented by research into working-class women (Reay et al. 2009). Yet she was passionate about helping others and fighting injustice and so slowly over several years came to champion the rights of neighbours helping them to pursue issues they had with various welfare benefits and employers. These interests and growing confidence led her to feel that she could return to education. Yet now she finds herself in the situation where her neighbours and many friends see her as a university student pursuing a middle-class route into a career—she no longer fits in, in her working-class neighbourhood, as she is seen as 'posh'. But at the same time, her age, how she talks and dresses marks her out as being working class to her university peers. Katherine then talks of being stuck in some sort of cultural 'no man's land'—as someone who has left her working-class neighbours and friends behind but yet is not accepted by the middle-class world of university.

> People say if you have a degree they listen to you, I went 'yeah, I know but why?' Why is it that you've got to have that bit of paper? So I went to show everyone that I can do it … I can get a degree the same as all the other people with money … I can do it, what makes me different from everyone else, why, because I am from Ashington [a well known former mining community] … I feel lonely, I don't fit in anywhere. I'm technically in the middle of two classes. I'm trying to fight for them [her working class neighbours] by fighting against them. The only thing is they think I'm selling out to them… 'Oh she's got money now because I've got a university coat and bag. Which in unfortunate as here [at university] I'm getting called, "you scruff". Here it's all new clothes and labels … I look and go, 'I'm, not you'. They drive around in their minis and their brand new cars. (Katherine)

Katherine offered an interesting insight about her class upbringing and the difficulties of being happy when life leads to transitions from one class to another. She said that originally she had become interested in university as she thought that a degree would help her to support her neighbours in the fights with government departments over their benefit claims. Yet paradoxically, becoming a student was seen by the people she knew as a sign that she was leaving them and the working-class behind. Katherine mused that to be successful in Britain today one has to leave one's past behind—one cannot go to university and have a career and still be working class. Consequently, the past year had been a difficult one for Katherine as she felt deeply uneasy about leaving her old working-class identity behind as she was so unsure about her future as a middle-class graduate.

Barry and Carl: Happiness and Alternative Lifestyles

The men in the Thirty Something sample, Barry from a middle-class background and Carl who grew up working class, share some of the features we see in the women. Notably, how material conditions (such as jobs and income) are important to wellbeing but there are also other sources of happiness. Admittedly, Barry and Carl are not representative of Thirty Something men in Britain today but the way that cultural activities, values and beliefs inform their wellbeing raises important questions about the more traditional economistic models used to explain wellbeing. They both appeared very happy with life despite living modestly. Only Barry completed the wellbeing questionnaire which supported his claims of being contented with life—scoring between 9 and 10 for satisfaction with key aspects of his life such as work, leisure, home, main relationship and family. Interestingly, he scored nature more highly and material possessions much lower than other interviewees. Their accounts pose questions for existing research in youth studies (Robb 2007) that traditionally view poor jobs and economic marginality as a key source of poor wellbeing. Barry and Carl's biographies instead suggest that wellbeing is a much more complex phenomenon than that depicted by many studies of young people.

When comparing the early experiences of Barry and Carl we see some of the differences that class can make to cultural experiences and life chances. Carl grew up in a large family (he had five sisters), his parents divorced when young and so money was tight in a lone parent household. Carl's mother supported him as best she could but she worked long hours in an administrative job. He passed a few GCSEs when 16 and worked in a series of clerical positions in local companies. He was fortunate that his sister worked in a local recruitment agency and helped him with employment throughout his teens and twenties. Carl says he would have preferred an apprenticeship in a more traditional industry but these were scarce as manufacturing was in decline in Northern England. Many researchers have documented the transformation of labour market opportunities for young men in former industrial areas (Nayak 2006; Jones 2002; Furlong and Cartmel 2004) and the emergence of a 'crisis in masculinity'. Many jobs now tend to be in 'feminised' areas such as office work and these tend not to offer the status, income or skill development that were seen in earlier male craft employment. One might expect that young men like Carl who are denied the opportunity to 'get on' through work may suffer from low self-esteem and poor wellbeing but this has not been the case. Although Carl admits he would have preferred a traditional factory job in manufacturing he also spoke of having a happy childhood and enjoying life despite working in low paid clerical positions.

Barry too had an unconventional life, despite well-educated parents they lived a peripatetic existence, moving around music festivals making a living selling homemade crafts. Barry was home schooled until 16 when he passed some GCSEs, then studied 'A' levels and attended university. He completed two years of a degree before dropping out of education to take up a series of temporary jobs in factories, shops and recently as a self-employed taxi driver.

As with Carl, Barry has lived most of his life at home with parents or in shared flats with friends surviving on a low income offered by temporary and casual jobs. Both have been unable to participate in the conventional consumerist lifestyles held up to be the norm for those entering their thirties. Designer labels, cars, overseas holidays and the like have all been beyond them yet neither spoke of the problems of being excluded from this way of life or of being unhappy with their lives. Their lives do seem rather different to the usual sociological portrayals that view young people's happiness as rooted in the ability to consume (Marcuse 2002; Furedi 2004; Davies 2015). To find a reason why both seemed content we have to examine their values and interests that influenced their 'alternative' lives.

The relatively good wellbeing of both men stemmed from a commitment to a way of living that nurtured them. For Carl, it was a passion for music that was at the heart of his life. From when he was a young teenager he had played instruments, sang and wrote songs with friends and this had continued throughout his life. Playing in a band and going to gigs was sociable—many of his friends had come through his love of music. Music was also important to his wellbeing as it allowed him to express himself—writing and performing songs about his life. He spoke of playing the guitar, singing and writing as a 'craft' and worked at improving his craftsmanship (Sennett 2009). Creating this music then was satisfying and rewarding and CDs they made had produced a small income for the band.

> It's that feeling when a song comes together and everybody has their input and then all those weeks of practising come together and then you perform it in front of people ... It's a massive passion. It's that end result, it's that feeling, it's amazing. When you are playing it it's indescribable. (Carl)

We see here again the role of flow experiences for good wellbeing and how nourishing activities such as music making can be for people's happiness (Csikszentmihalyi 2002). Marx (1983) too would recognise the significance of these activities that allow people to control a productive process that collectively and creatively produces goods that have use and exchange value.

In contrast, Barry's life was shaped much more by socialist politics and his beliefs in environmentalism. He attended activities that socialised him into an alternative way of living such as socialist rallies calling for workers' rights and protests against environmental destruction. Barry and his parents were part of local networks of activists and they shared their skills in recycling, gardening and DIY projects so they could be self-reliant and challenge traditional lifestyles and the reliance on waged work and consumerism. A balanced life was integral to how he and his parents had lived—they enjoyed a lifestyle that eluded the women in the 'Thirty Somethings' sample. Like Carl, Barry also spoke of getting much from his alternative lifestyle that allowed him to be content. He had a wide circle of like-minded friends who were socialists and environmentalist and they worked together on local campaigns. He had developed numerous skills over the years such as motor mechanics that meant he could repair cars, help others and make some money. His beliefs in socialism (and writers such Gorz 1985) also gave him a rationale for his alternative way of life—to make sense of how work could be exploitative and a corrosive consumerism underpinned capitalism. His work as a taxi driver was just a means to an end—to make enough to live whilst allowing time for all the other activities that he valued and which nourished him.

Barry and Carl: Settling Down and Being Happy?

For most of their twenties both Barry and Carl had enjoyed carefree lives, with few responsibilities, living cheaply at home, only working around their interests and hobbies. Neither were particularly interested in pursuing careers through their work, investing their identities and seeking their wellbeing elsewhere. Yet as they approached their 30th birthdays both had embarked on serious relationships, setting up home with their long-term girlfriends. Gradually, they were beginning to recast their lives—moving from a life of single males to a shared life as part of a couple. Both spoke of the fun and excitement of setting up home but also the challenges of these changes and impact on their wellbeing.

Moving into a shared home for both men was seen as a positive step that would enhance their wellbeing. Both were conforming to traditional ways of living a happy life in the UK—sharing a home with a long-term partner, buying a house, having children and getting married (Ahmed 2010). Yet these changes do seem surprising as Barry and Carl had both lived a bohemian existence for many years, flitting from one temporary job to another, concerned more with their music or politics than anything else. How are they able to stay true to themselves, to hold on to those values and interests that make them who they are whilst also embracing a more conventional consumerist way of life? Surely setting up a home, planning weddings and a family would threaten these core principles they valued and with it undermine their wellbeing?

> Even though I have a son and a great girlfriend there is something missing. It is like Groundhog Day, it is a bit repetitive. You know, you get up for work, have breakfast with my little boy Thomas and Ann [girlfriend] goes off to work. Then come home, have your tea, then its bath time, then bedtime for Thomas. Then you just have an hour and a bit to yourself. I know that's life but I mentioned it to Ann that we should try and do more of a night … I always feel that we need to be doing things all the time. I don't like just sitting around. (Carl)

Although Carl anticipated these changes (his friends were settling down, getting married and becoming parents) he nevertheless spoke about how there were difficult compromises and sacrifices he had to make. A key feature of the happiness maps we drew in interviews was the difficult transition from life as a single man with free time to one of a partnership in a shared home. Keeping their small flat clean and tidy, organising the cooking of meals and washing of clothes all had to be discussed and negotiated. Carl admitted to being overly concerned about the flat being tidy and well organised and this had led to arguments and disagreements with Ann his girlfriend who was much more relaxed about tidiness. The arrival of their son, Thomas added greatly to the clutter around their flat and Carl spoke of his dismay on arriving home from work to find a chaotic assortment of babies' clothes, food and toys littered around their home.

In interviews we discussed together the riddle of happiness and a good life—that he had seen settling down as a positive move as he loved his girlfriend, yet there were also aspects of his new life that were a source of unhappiness. Over several months Carl said that he had come to terms with the messiness of their flat—compromising with his girlfriend—she made an effort to tidy up and he tried to be relaxed about mess around the home. However, there were features of their life together that were more difficult to manage and posed a more serious threat to his wellbeing. The arrival of their baby had made them realise that their flat was too small for them but their search for a small house and affordable mortgage was hampered by their poorly paid jobs. Carl spoke of being stuck in rented accommodation, unable to afford the move to a better more spacious family home. His series of poorly paid jobs over the past 10 years meant he had little savings for a deposit and a number of mortgage companies saw him as a risky borrower. As we saw in the previous chapter with the younger interviewees, Carl's story is a familiar one for many young people in Britain who struggle to find accommodation they can afford as they are caught between rising housing costs and poorly paid employment (Willetts 2010; Walker and Jeraj 2016). These social constraints on young people's lives—the so-called 'generation rent' undermined efforts to live well and threatened wellbeing. Unlike in the past, when Carl could be relaxed about work and money he realised he would have to take work far more seriously as he needed a bigger salary to secure a mortgage. In many ways Carl's unease about his life and his concerns about his wellbeing stem from how he feels he is leaving behind a life he loved—a relaxed attitude about waged work, time with his friends and time for making music. When we spoke about the future he hoped he might be able to combine these earlier ways of living with what he has now with his partner—a good balance between the old and new Carl.

In Barry's case his transition to a long-term relationship and a shared home appears to be more positive for his wellbeing than Carl's experiences. The mapping poster we did together only pointed to a few aspects of the recent changes to his life that were problematic. He had known his girlfriend (who was a primary school teacher) for three years and they had bought a house together and were planning their wedding. It was the costs around the house purchase and impending wedding that posed some concerns for Barry. Like Carl, Barry had always valued the freedom to work part-time or take time off to pursue his own interests—usually around political and environmental campaigns. As a socialist he also felt strongly that many jobs were exploitative, reducing workers to wage slaves. The challenge for Barry was to earn enough to help pay for their new lifestyle without resorting to the low paid, exploitative jobs he had done in the past. Becoming a self-employed taxi driver offered a good solution to this dilemma, as he was able to earn a living with flexible hours and some autonomy.

The shift to a more consumerist way of living that posed issues for both of our young men seems to have been managed rather better by Barry than Carl who was tied into a more traditional bureaucratic and at times disempowering work environment. Barry's transition to shared living arrangements was also more successful as he and his girlfriend shared more in common than Carl and his partner. Where Carl's girlfriend tolerated his interest in music and his ambivalent attitude to waged work, Barry's partner shared his political views and interests in environmentalism. Barry also admitted they were both disinterested in being tidy and so over the past year had slowly agreed on how best to manage their home together—the cooking, cleaning and shopping. When asked about his wellbeing, Barry said that these were the 'happiest times of his life'.

The 'Thirty Somethings' wellbeing differed from the young people in Chap. 7 as they were 'settling down' and becoming parents, which posed some difficult choices for them. We have a sense here of wellbeing as a practical activity as the interviewees were reflecting on how best to juggle different responsibilities, interests and pursuits. Although traditional survey research documents the sources of women's wellbeing these snapshot approaches seldom show how happiness functions across particular biographies and in specific contexts. Whereas a biographical approach allowed us to explore how different life course events had shaped the women's identities and in turn their wellbeing. Although they all had some control in these matters, class and gender relationships influenced their resources and opportunities—Jenny and Katherine both struggled to live well as they had a modest start in life and their relationships with partners were sometimes problematic for their wellbeing. Sally in comparison had more success professionally yet she too was struggling to combine a career with her hopes for marriage and children. The two men in the sample were unusual in that their alternative lifestyles, rooted in interests and hobbies supported good wellbeing despite their meagre incomes—a finding at odds with the usual consumerist notions of young people. Both, however, have recently set up homes with partners and were dealing with change that illustrated how happiness is bound up with trade-offs, negotiations and conflict with family, friends and lovers. As people age these relationships and efforts to be happy can become even more complex and challenging as we see in the next chapter—happiness in mid life.

Bibliography

Ackerman, N., & Paulucci, B. (1983). Objective and subjective income adequacy: Their relationship to perceived life quality indicators. *Social Indicators Research, 12*(1), 25–48.

Age UK. (2015). *Combating loneliness*. London: Age UK. Retrieved from http://www.ageuk.org.uk/health-wellbeing/relationships-and-family/befriending-services-combating-loneliness/

Ahmed, S. (2010). *The promise of happiness*. London: Duke University Press.

Aknin, L., Norton, M., & Dunn, E. (2009). From wealth to wellbeing: Money matters but less than people think. *The Journal of Positive Psychology, 4*(6), 523–527.

Aristotle. (2009). *Nicomachean ethics*. Oxford: Oxford University Press.

Bourdieu, P., & Passeron, J. C. (1990). *Reproduction in education, society and culture*. London: Sage Press.

Cederstrom, C., & Spicer, A. (2015). *The wellness syndrome*. London: Polity.

Crompton, R., & Lyonette, C. (2006). Work life balance in Europe. *Acta Sociologica, 49*(4), 379–393.

Csikszentmihalyi, M. (2002). *Flow: The classic work on how to achieve happiness*. London: Harper and Row.

Davies, W. (2015). *The happiness industry: How the government and big business sold us wellbeing*. London: Verso.

Deary, V. (2014). *How to live*. London: Allen Lane.

Dolan, P. (2015). *Happiness by design: Finding pleasure and purpose in everyday life*. London: Penguin.

Durkheim, E. (1984). *Suicide*. Harmondsworth: Penguin.

Durkheim, E. (2014). *The division of labour in society*. New York: Free Press.

Esping-Anderson, G. (2009). *Incomplete revolution: Adapting welfare states to women's new roles*. London: Polity.

Furedi, F. (2004). *Therapy culture: Cultivating vulnerability in an uncertain age*. London: Routledge.

Furlong, A., & Cartmel, F. (2004). *Vulnerable young men in fragile labour markets*. York: York Publishing.

Gilbert, P. (Ed.) (2005). *Compassion: Conceptualisations, research and use in psychotherapy*. London: Routledge.

Goffman, E. (1956). *The presentation of self in everyday life*. Edinburgh: Edinburgh University Press.

Gorz, A. (1985). *Paths to paradise: On the liberation from work*. London: Pluto Press.

Guardian. (2014, October 14). *Social care is on the cusp of a crisis.* Retrieved from http://www.theguardian.com/social-care-network/2015/oct/14/social-care-cusp-crisis

Hochschild, A. (2003). *The managed heart: The commercialisation of human feeling* (2nd ed.). Berkeley: University of California Press.

Hockey, J., & James, A. (2003). *Social identities across the life course.* London: Palgrave.

Hyman, L. (2014). *Happiness: Understandings, narratives and discourses.* London: Palgrave.

Jahoda, M. (1982). *Employment and unemployment: A psycho-social analysis.* Cambridge: CUP.

Jones, G. (2002). *The youth divide: Diverging paths to adulthood.* York: Joseph Rowntree Foundation.

Kahneman, D. (2011). *Thinking fast and slow.* London: Penguin Books.

Layard, R. (2005). *Happiness: Lessons from a new science.* London: Penguin.

Lunau, T., Bambra, C., Eikemo, T., Van der Wel, K., & Dragano, N. (2014). A balancing act? Work-life balance, health and well-being in European welfare states. *European Journal of Public Health, 24*(3), 422–427.

Marcuse, H. (2002). *One dimensional man: Studies in the ideology of advanced industrial society.* London: Routldege.

Marx, K. (1983). Alienated labour. In E. Kamenka (Ed.), *The portable Karl Marx.* New York: Penguin.

Mills, C. W. (1959). *The sociological imagination.* Oxford: Oxford University Press.

Nayak, A. (2006). Displaced masculinities: Chavs, youth and class in the post-industrial city. *Sociology, 40*(5), 813–831.

Nussbaum, M., & Sen, A. (1993). Introduction. In M. Nussbaum & A. Sen (Eds.), *The quality of life.* Oxford: Clarendon Press.

OECD. (2002). *Babies and bosses—Reconciling work and family life* (Vol. 1, Australia, Denmark and the Netherlands). Paris: OECD.

Pahl, R. (1995). *After success: Fin-de-Siecle anxiety and identity.* Cambridge: Polity Press.

Pahl, R. (1998). Friendly society. In I. Christie & L. Nash (Eds.), *The good society.* London: Demos. Retrieved from http://www.demos.co.uk/files/thegoodlife.pdf?1240939425

Putnam, R. D. (2001). *Bowling alone: The collapse and revival of American community.* London: Simon and Schuster.

Reay, D., Crozier, G., & Clayton, J. (2009). 'Fitting in or sticking out': Working class students in UK higher education. *British Education Research Journal.* IFirstarticle, Retrieved from http://dx.doi.org/10.1080/01411920902878925

Robb, M. (2007). Wellbeing. In M. J. Kehily (Ed.), *Understanding youth: Perspectives, identities and practices* (pp. 181–214). London: Sage/Open University Press.

Roberts, K. (2007). Work life balance—The sources of the contemporary problem and the probable outcomes. *Employee Relations, 29*(4), 334–351.

Sayer, A. (2011). *Why things matter to people: Social science, values and ethical life*. Cambridge: Cambridge University Press.

Schiller, F. (2004). *Letters on the aesthetical education of man*. New York: Dover Publications.

Sennett, R. (2009). *The craftsman*. London: Penguin.

Shaw, S. M., Andrey, J., & Johnson, L. C. (2003). The struggle for life balance: Work, family and leisure in the lives of women teleworkers. *World Leisure Journal, 45*(4), 15–29.

Smart, C. (2007). *Personal life*. Cambridge: Polity Press.

Tait, M., Badget, M. Y., & Baldwin, T. T. (1989). Job and life satisfaction: A re-evaluation of the strength of the relationship and gender effects of a function of the date of the study. *Journal of Applied psychology, 74*, 502–507.

Taylor, C. (1991). *The malaise of modernity*. Toronto: House of Anansi Press.

Thin, N. (2012). *Social happiness: Theory into policy and practice*. Bristol: Policy Press.

Toynbee, P., & Walker, D. (2015). *Cameron's Coup: How the Tories took Britain to the brink*. London: Guardian/Faber.

Walker, R., & Jeraj, S. (2016). *The rent trap: How we fell into it and how we can get out of it*. London: Pluto Press.

Wilkinson, R., & Pickett, K. (2009). *The spirit level: Why equality is better for everyone*. London: Penguin.

Williams, M., & Penman, D. (2011). *Mindfulness: A practical guide to finding peace in a frantic world*. London: Piatkus.

9

Happiness in Mid Life

In the previous chapters we witnessed the different ways that people of various ages have experienced happiness and how class and gender processes for example can influence wellbeing. A key aim of the book is to use people's narratives to offer a more nuanced analysis of happiness than we see in many wellbeing surveys. By using biographical interviews we witness how notions of happiness and living well are woven into the difficult choices and trade-offs people make in families, work and leisure. As we have already seen with the young people sample and the 'Thirty Somethings' people have to balance the need for an income with other interests and responsibilities if they are to live well. As one gets older and one's responsibilities increase, does it become more difficult to be happy in middle age? As people move through life they seek to flourish as individuals but they also want many other people, such as lovers, children and friends to be happy too. These hopes for happiness are fraught with difficulties, as they are ambitious but also as people change with age, so do their ideas of happiness. As people approach mid life after decades of familiar routines how do they continue to live well? Are some people more vulnerable to unhappiness as they get older, or are there skills some

M. Cieslik, *The Happiness Riddle and the Quest for a Good Life*,
DOI 10.1057/978-1-137-31882-4_9

people develop that support their wellbeing? These are some of the questions I posed to the middle-aged people I interviewed for this project.

The mid-life sample was comprised of five people, two of them could be classified as working class—Kevin, a self-employed gardener (aged 42 when interviewed in 2012), and Paula who was 48 years old when interviewed in 2014 and had been claiming disability benefit for several years. The middle-class interviewees comprised James who was a partner and owner of a medium-sized business (aged 55 in 2011); Nigel who was educated to postgraduate level and was an activist and part-time lecturer (aged 47 in 2011); and Kate who had attended university and worked as a teaching assistant (aged 54 when interviewed in 2011). When comparing their views on their personal wellbeing (informed by questionnaire responses, diary entries and interview data) it was apparent there were some significant differences in how happy they had felt in recent weeks and months. James who was the most affluent and successful of the five spoke of being mostly unhappy in recent months as he had recently separated from his wife. Whereas Paula scored very highly on the wellbeing questions as she had recently began a new relationship she said had brought her much joy. Both Nigel and Kate are unusual in having a university education yet have pursued relatively 'alternative lives' mixing employment with volunteering and activism. They both spoke of struggling on low incomes but overall they were mostly happy—far more so than the more affluent James. Even Kevin, who worked long hours for poor pay, was happier than James, speaking of how his long-term relationship, friends and leisure interests sustained his wellbeing.

The findings from the data runs counter to the conventional arguments offered by sociologists and wellbeing researchers who use quantitative surveys. This existing work suggests that social class background (understood via the incidence of poverty and inequality) greatly influences life chances so that the affluent middle and upper classes with higher incomes, better education and housing are more likely to be contented than the less affluent (Children's Society 2015; Layard and Dunn 2009; Wilkinson and Pickett 2009). Many economists (Stevenson and Wolfers 2008; Kahneman and Deaton 2010) and a few sociologists such as Hagerty and Veenhoven (2003) also point to survey data that correlates higher income with increased life satisfaction. So how then do we explain these case studies that suggest that material resources and opportunities are not always associated with better wellbeing? We could argue that such findings are an artefact of the small, unrepresentative sample used in this research and that larger samples would produce more predictable results. However this chapter illustrates as do other chapter in this book that by analysing the life course of different individuals we can see the cumulative effects of aging on individuals and how this impacts on how they understand and experience happiness. Respondents' subjectivities or their sense of self became more complex as they aged and so the meanings around the notion of a good life became more nuanced compared to their younger selves (see Jenkins 1996: 21; Hockey and James 2003: 201; Hatch et al. 2007). In particular, aspects of earlier life events (and with it certain desires or fears) can be internalised shaping people's psychologies that influence wellbeing in often unexpected ways. This is the issue of fallibility and how people have partial insights into their circumstances and hence struggle to make informed choices about how best to live (Kahneman 2011; Dolan 2015; Gilbert 2006). Hence trying to live well can often seem like a confusing riddle. Wellbeing research that relies on relatively simple life satisfaction/affect questionnaires abstract from these complexities and therefore neglect these issues of meaning and the significance they have for people's understanding and experiences of happiness.

Another reason why the data offered here runs counter to the conventional findings on income and wellbeing is that I explore happiness as a social process rather than an individual characteristic. As we see in this book, my participants documented how their wellbeing evolved out of a myriad of different social relationships over many years. So their effort to live well involved struggle, negotiation and conflict over what happiness meant to them and their family and friends. I draw on some sociological concepts such as emotional labour, habitus and discourse to show how power and resources do have a bearing on these processes by which people struggle to be happy. Yet as we have seen in earlier chapters these structural features of people's lives, such as their income, work experiences or class cultural background do not always determine their wellbeing. The following biographies illustrate instead that individuals, despite the challenges posed by their fallibility, creatively adapt and manage their wellbeing that generates complex relationships between structural characteristics of individuals and their subjective wellbeing.

James: Middle Class and Happy, but a Mid-life Crisis?

At first sight James' biography confirms the conventional wisdom on how best to be happy. His educated parents encouraged him to work hard at school and attend university. Sociologists would tell us that middle-class families have a distinctive set of practices (or habitus) that confer cultural, social and material resources that in turn shape the ability to compete for opportunities (in a structure or field) that can aid the successful route through education and into a professional career (Bourdieu and Passeron 1990; Bourdieu 1977). He graduated from a top university where he met his wife to be who was also middle class. They set up home and secured well-paid professional careers, buying an expensive house and enjoying holidays and expensive cars and buying private education for their children. Nevertheless, one way in which survey research acknowledges that there is a complex relationship between income/social class and wellbeing is by suggesting that many people who appear successful do experience a temporary mid-life crisis in the fourth or fifth decade of life (Blanchflower and Oswald 2008). Commentators do stress that such experiences are temporary—with the years before and after such a crisis reaffirming how affluence and higher income are correlated with good wellbeing.

It is James who best conforms to this image of the middle-aged man who, despite the trappings of a successful life, had become disenchanted, triggering an existential crisis and efforts to remake his life. He had separated from his wife of 25 years, 18 months before I first interviewed him. Since then he has set up home in a small flat, had relationships with other women and established the sort of life more commonly associated with younger, single men—nights out at pubs and clubs, working out at the gym and so on. He has re-evaluated his relationship with work—now working four days a week so he has more leisure time. Yet, despite these changes to his routines he still said that he often felt unhappy. He regretted the way his marriage had ended and the difficulties he has seeing his children. He also missed the companionship and family home he once had—he felt lonely coming home from work to an empty flat. The financial costs of divorce were also a concern—setting up another home and supporting his wife and family were very expensive and ended his hopes of early retirement.

James's experiences illustrate the sorts of trends captured in large-scale surveys of happiness and aging that show that many middle-aged people experience a pronounced dip in their wellbeing (ONS 2016). Yet by examining his biographical profile we see the development of this crisis evolved over many years rather than it being associated just with mid life. So the idea of upheaval in mid life does simplify the complexity of people's lives and reflects how surveys tend to offer snap shots that neglect processes that influence wellbeing. Although objectively James was successful in achieving the many goals associated with a middle-class life we see that subjectively his wellbeing had been problematic for many years. This is a finding that questions the usual relationships between affluence and wellbeing offered by survey research.

James had experienced bouts of depression in the years before interview and attended counselling sessions where he had tried to make sense of the breakdown of his marriage and his poor wellbeing. Although it was only recently he had left his wife, he said that on reflection there were signs, when he was in his 30s that he was experiencing some mid-life anxieties and problems with his marriage. He acknowledged that with hindsight he could now see the fault lines in his marriage but these were difficult to see at the time. This is a good example of fallibility and how people when caught up in events struggle to assess how best to live and be happy (see Kahneman 2011; Dolan 2015).

> My brother-in-law had a nervous breakdown, quite a few years ago and, oh this must be about maybe 14, 15 years ago. I was talking to him and I said 'You know I don't feel really happy with my life, you know, I feel like there's something missing' and he says 'Oh I've got a friend up in Newcastle who's a psychiatrist go and see her'. So I went to see her, I didn't tell my wife, and the psychiatrist was really good actually … There was an emptiness in my life and I did use to read these self-help books and that really annoyed my wife as well. She used to think, 'well, you're reading these books because you're not happy with me'.

Research into mid life suggests it can be very unsettling to feel that the important choices that one has made in life (about jobs, partners and homes) no longer appear to be the right ones (Levinson 1991). When the key features of life inexplicably go from being a source of happiness to a source of sadness and disappointment one is left searching for some coherent and logical explanation for this transformation of one's wellbeing. James has spent many hours over many years struggling to make sense of these changes to his life. At first, the inability to make sense of these changes contributed to his poor wellbeing—there didn't seem to be a logical or rational explanation for what had happened. With time, however, he now feels that he has been able to begin to construct some coherent and logical account of the breakdown of his marriage that has become part of the process of moving forward with his life. He traces some of the problems with his marriage to a confluence of events over several years in his thirties and forties. He had become disenchanted with his very busy job and his youngest child had just started school so as a couple, James and his wife had more time together. But instead of this extra time being a good opportunity for them to enjoy it revealed fault lines in their relationship.

> When you have four children who are very close in age you tended to just get through life, you know, you just had to, it was a struggle, it was just one nappy change after another and you didn't really have time to have a relationship with your wife …you suddenly thought, well is this it, you know, is this it for the next 20 years or something, am I still going to go to that office where I have been for the last 20 years, is this going to be the same, and I suddenly thought, it was a mid-life crisis I suppose, this is not for me. And I had this idea that I wanted to go to Australia. I've met a few Australians and I've always really liked them and I detest the climate in the North East and I'd really had this obsession with going to Australia and, so I persuaded my wife, who wasn't keen, I mean she likes travelling, but she wasn't keen on the idea of emigrating, to go there for three months … And we had a great time, a fantastic time, but I really wanted to move out there and she didn't.

Though they were discussing how best to change their lives so they could both be happy they had very different views on how to do this. There are similarities here with the women in the previous chapter that illustrate how happiness can be infused with tensions and conflicts. James wanted a change of career, a different climate and routines that Australia could offer whilst his wife favoured only minor changes to their old jobs, home and routines. They both tried to negotiate a way out these disagreements. Hence in the years following their trip to Australia there were various attempts to fashion a more enjoyable, happier lifestyle with long family holidays abroad and the purchase of a holiday cottage in Spain. These efforts to patch up their lives however did not seem to work as they hoped. The interviews with James reveals how, with hindsight, he now sees this period in his life as one marked by a shift in his relationship with his wife—growing differences between them about how to live which with time came to be a source of sadness, threatening their once strong and nourishing marriage.

'Keeping Up Appearances': Fear of a Failing Marriage

James had spent much time mulling over the reasons for the failure of his marriage and he suggested that these early disagreements about how to live together marked the beginning of the gradual unravelling of his relationship with his wife. Interestingly, James remarked how at the time neither he nor his wife really acknowledged (to themselves or to each other) the tensions in their marriage. Like much research on family breakdown (Roberts et al. 2009), James suggested that perhaps it was the busy routines and pressures of work and raising small children that obscured some of the underlying problems they had. Certainly these played their part in obscuring the choices they faced. There was also a sense they were not communicating honestly with each other about their concerns about their marriage. We see here how ideas about happiness and a good life become woven into not only the fabric of a relationship but also the psyche of the individuals. If we draw on some psychoanalytical ideas we can suggest that both James and his wife were fearful of these fault lines in their marriage—denying, repressing or rationalising them—throwing themselves instead into other activities such as rearing the children, success at work and planning lavish holidays (Freud 1992). Their inability to face their difficulties—their fallibility was partly rooted in the way that fears and desires play out in our psychic selves.

> That's my biggest regret really that she didn't come to me before. I mean I've gone over this a million times in my head and with her as well, she must have been unhappy with me for a number years and she said 'Well I was, but you know, I thought it was just a passing phase, it would just go away'. But it wasn't ... But I never had the courage to leave my wife and she had never had the courage to leave me ... I think it was partly that and partly probably her fear of being alone, my fear of being alone, financial fear, a lot of things. It was just easier just to continue the relationship, it was almost like a platonic one, but that, it really wasn't for me, I didn't like it. (James)

This process of how people can actively avoid or obscure underlying anxieties, desires or threats that are significant for their happiness does pose serious challenges for wellbeing research. It raises questions about the utility of satisfaction surveys when questions need to be posed in much more subtle and specific ways to capture the levels of meaning associated with these life events. Even if we use qualitative methods to explore wellbeing it raises questions about the knowledge that people have about their wellbeing and its sources. Sometimes people are unsure about why it is they may feel happy or sad as such emotions arise out of complex features of their psyche which is turn are linked to some distant, often forgotten events (Hatch et al. 2007). When being interviewed we edit out much detail about our pasts, we also forget more than we remember and see coherency and causal links where there is chance or luck (Kahneman 2011) and there are differences between narrating a life as we look back on events in interview and how we actually lived at the time.

A further complex issue we see in James's wellbeing is the sensitivity that he and his wife felt about displays of happiness to each other. This illustrates another way that happiness functions, as a 'form of performance for others' (Goffman 1956) and which contributed to the difficulties James faced trying to sustain a happy life with his wife and children. He spoke of being conflicted between wanting to be honest about his unhappiness yet not wanting to blame his wife for this situation. He did not want to suggest that she was the main source of his unhappiness as his sadness was more complicated than this. At first he was quite open with his wife about changing their lives and being happier. But this just added to their woes.

> She really felt that me wanting to go to Australia was that I didn't want her anymore, wasn't happy with my lot, I wanted something new and I wasn't happy with her. Which wasn't the case at all, it was a case of I wanted to start a new life for the family in Australia, it's going to be a fantastic thing, you know, it would be like we've done one half of our life here and now the second half of it was going to be something completely different, but she didn't quite understand that. (James)

For many years it appears that James and his wife did not admit to each other or even to themselves that there were serious problems with their marriage. James's wife saw his interest in self-help books and holidays as a criticism of her so he learnt to keep these things to himself. He spent several years 'acting' as if his marriage was fine—performing outwardly the role of a happy husband and father—as well as internally and emotionally working at being a loving member of the family—what the sociologist Arlie Hochschild (2003) refers to as 'deep and surface acting'. Here we see the way that happiness is something that people do, not just for themselves, but for others too rather than the popular idea of happiness as a characteristic of an individual. Rather here, it is a form of practical, social activity that also involves thinking through the needs of others and offering kindness, compassion and care. Although for James there was a dissonance between his inner self (his hopes for a good life—being loved by his wife and a new job) and his exterior 'happy self' his performances over many years were mostly successful in sustaining his place in his family. So much so that he was surprised when his wife finally admitted their relationship had run its course.

Kevin and Paula: 'Life's Been a Struggle But Happy'

In marked contrast to James's biography, Kevin and Paula had enjoyed much less parental support when young because of their class and gender backgrounds, pursuing more insecure routes into adulthood with poorer paid jobs, leisure and housing. Yet both suggested they had been happy and contented—far more so than James. Disadvantaged individuals that one might expect to have poor wellbeing it seems can often flourish because of efforts they make to lead a balanced, nourishing life that is neglected by the more career-focused middle classes. Consequently, Kevin and Paula, unlike James did not understand their lives in terms of a mid-life crisis. For them, life had been much more of a life long struggle to live well. Such findings confound wellbeing research that claims that affluence correlates with better wellbeing (Stevenson and Wolfers 2012) or that wellbeing might dip in mid life (Blanchflower and Oswald 2008). These biographies also contrast with the conventional sociological wisdom that social disadvantage is often associated with suffering and poor wellbeing (Wilkinson 2005: 48).

Gardening and Cycling: Happiness Amidst Economic Decline

When interviewed Kevin spoke of mostly being contented with his life (which was confirmed by his responses to the wellbeing questionnaire), although at times he had experienced some low points because of difficulties finding work. Like many men of his age in the North East of England he has experienced the effects of declining industry and loss of well-paid jobs (Webster et al. 2004). The biographical timeline we constructed charted his education to employment transitions, moving from school into a series of low paid, low skill jobs (such as school caretaker and insurance salesman) and experiencing periodic redundancy as companies closed. In his twenties, after a short training course he became a self-employed gardener, subsidising this income with a part-time job for the local authority, erecting shelters for weekend markets. However, after ten years he was made redundant by the council and was left to work longer hours as a gardener to make up the shortfall in his income, which was difficult, particularly in the winter months when work was scarce.

Commentators suggest that the economic decline and growth in poverty in many parts of Britain since the 1980s has impacted significantly on the wellbeing of citizens (Wilkinson and Pickett 2009; Mack and

Lansley 2015)—certainly Kevin suggested that life had been a struggle at times. Men of Kevin's age, who left school in the 1980s, had been socialised into earlier, more traditional cultural assumptions about the availability of well-paid working class jobs yet found themselves dealing with the collapse in these once plentiful working class careers. The mismatch between aspirations and opportunities led many researchers to talk of a crisis for these communities (Brown 1987; Weis 1991). For many working class men in particular, researchers pointed to a 'crisis of masculinity' as they struggled to achieve the status and identity that secure jobs had offered them in the past (Mac an Ghaill 1996; Roberts 2014). Consequently, these processes of economic decline are associated with data that points to growth in the numbers with poor mental and physical health in the former industrial areas of Britain (Dorling et al. 2007).

> I'm not sure if its depression but you get sick, don't you and you just have to get on with it. I was always depressed if I had no money. That's what depressed me, if I had no money, but once you get a bit of money it doesn't depress you. (Kevin)

Despite the economic insecurity that has characterised his life Kevin feels that he had been lucky, when compared to his working class peers, many of whom were also finding it difficult to secure work and were often unemployed. This data illustrates not only how we compare ourselves to others (social comparisons with reference groups) when making judgments but also the significance of cultural and personal expectations for a standard of living and the happiness it is seen to bring (Layard 2005: 47). Although aspiring to a secure job, Kevin's working class habitus had also imbued him with rather more modest ambitions about the quality of work, 'success' and lifestyles than those from more affluent social backgrounds. Such experiences reflect social processes that Durkheim also noted in the nineteenth century where the ability to manage or discipline ones expectations can help one to be happy even in difficult times (2014). Kevin's more modest hopes for the future have surely been significant in accounting for his relatively good wellbeing in recent years. But there are other factors such as his access to various resources that appear significant for his life course transitions and wellbeing.

> My mother's boyfriend worked on the market stalls and when he retired I got the job. It's 'who you know in this world, not what you know', he put a word in for us and I went to see the bloke. When I got that job on the council it set me on in my life. I did it for 10-12 years and it was £150 a week for only 4-5 hours' work. It helped me get a mortgage as I had another job [as a gardener] … that was £150 as well. That was £300 a week, which was a lot of money twenty years ago. (Kevin)

At this time, Kevin had met his future long-term partner and so her additional income helped them to set up home and achieve some independence from his mother. Having the resources to set up home also contributed significantly to his positive wellbeing (see Easterlin 2003)—something that he says was denied to many of his peers at the time. Kevin also felt lucky as he had profited from the social capital embedded in family networks in the local area. These contacts helped him later on when establishing his gardening business as friends and family recommended him to clients.

In interviews, however, Kevin did point to other reasons why he had good wellbeing despite the economic insecurities he experienced. This is one of the challenges of researching wellbeing as happiness and a good life have many sources that operate through different social domains. Whereas traditionally researchers tend to explore wellbeing in specific contexts like families, work or health that reflect existing academic specialisms. Kevin had managed to sustain his relationship with his partner despite knowing each other for 20 years, they had common interests and still enjoyed spending time with each other (Layard 2005: 65). Kevin also had a rather more balanced life than James, combining his waged work with other interests, notably around sport and his friends he met through these activities. In particular Kevin had grown up in a family of cyclists—his father and uncle introducing him to the local club when he was child.

> When I was younger your life was the bike, Saturday, Sunday, all the week … and then we'd go out drinking the Friday night and then the Sunday night … The bike is a good release. You go on the bike and have a ride and then you forgot about it, don't you when you go on your bike. It's freedom, isn't it? You don't worry about your bills when you are on your bike. (Kevin)

Ever since he has continued to ride his bike two or three times a week, socialising with club members on the weekend. Kevin suggested that his lifelong interest in cycling had played a key part in keeping him happy and contented. As we have seen in earlier chapters (with Marx's notion of flourishing), the opportunity to engage in physical exercise, to experience the beauty of nature and chance to share these experiences are all linked to positive wellbeing (Lunau et al. 2014; Joseph Sirgy and Wu 2009; Seligman 2002).

> I like time trialling because you can try and beat your best time, your P.B ... It's really hard. But you got to do it, haven't you? Once or twice when I really try, it's really hard, but you do it, don't you and afterwards you feel champion ... Then at the finish when you are all talking about your times, it's brilliant isn't it? You do get addicted to it because you want to improve ... People said to me about a doing a 12 hour race, I said right I'll prove it to them fuckers, I'll do it, so I did a 12 hour. Did 227 miles with just 5 minutes off the bike and I knew I could do it and I was determined to prove them wrong. So I rode a 12 hour. I'd never ride another one! (Kevin)

For some Marxist sociologists the involvement of the working classes in leisure and sport allows a brief respite from the alienation of work, ensuring their continued commitment to employment that ultimately undermines their wellbeing (Marcuse 2002; Adorno 1941). But in Kevin's case his involvement in cycling was far more significant for his wellbeing than these traditional analyses suggest. Being born into a cycling family the sport had been the basis of much of his social life that extended through a complex network of friends in the cycling community in the North East of England. Over 40 or so years his life and his identity had been shaped by the many rituals of cycling—the Sunday club runs, monthly club meetings, cycling holidays, watching the Tour de France, training, racing and nights out (see Bacon 2015; Barter 2013). Kevin had perhaps fared better than many others dealing with economic decline as the cycling had offered a range of valuable resources—many friends and their social capital and also a status, self-worth and meaning through his sport.

Csikszentmihalyi's notion of 'Flow' is also significant here for listening to Kevin we can hear how cycling involves working on the self to positively shape his wellbeing (Csikszentmihalyi 2002). In the interview Kevin produced many photographs, newspaper cuttings and dairies that charted his involvement in cycling over four decades. This material documented year by year the development of his fitness, times and speeds for races and different training regimes. He had become accustomed to challenging himself physically and mentally when riding, particularly when racing and training. This cycling had all the hallmarks of flow events—being immersed in a challenging activity, being in control, yet lost in the bodily exertion fighting one's way towards the finish line. Then the relief, the buzz, a sense of achievement and also the endorphin rush after the race had finished.

Paula: 'A Lifelong Crisis But I've Been Lucky'

Paula, the other working-class member of the mid-life sample spoke of having good wellbeing in recent years, much better than the middle class James, despite her dependency on sickness benefits (she was diagnosed with multiple sclerosis at 26). This finding runs counter to much wellbeing research that links higher incomes to greater happiness (Hagerty and Veenhoven 2003; Stevenson and Wolfers 2008). Although over the longer-term, Paula's wellbeing had overall been poorer than James—but not necessarily because of differences in their material situation. Paula's biography illustrates some of the complexities and processes that underpin wellbeing that existing survey research often neglects. In particular, how a lifetime of experiences can shape expectations about a good life/wellbeing so that individuals have very different understandings of happiness that are missed by traditional survey techniques. Paula led a frugal, simple life, living with her partner in social housing—one that might be associated with poor wellbeing. Yet as she had experienced many years of abuse in a dysfunctional marriage but had now left this behind she spoke of feeling very happy, even though materially she had very little. Whereas James had a much more comfortable life, materially than Paula—what they did share was the way in which ones wellbeing can be progressively undermined by the failings of an intimate relationship—love that has gone wrong.

Paula's biography, like Kevin's, also raises questions about the mid-life crisis thesis. Like Kevin, Paula experienced a more disadvantaged route through childhood into adulthood. Is it because of these challenges and their responses to them that they avoided a later mid-life crisis? Because of their class and gendered backgrounds both have been confronted by difficult choices, unlike James—redundancy, managing on low incomes, securing affordable housing—so their biographies have been lifelong crises not just one at mid life. Paula's story illustrates the way that power relations work through social class and gender to frame biographies and wellbeing as noted by various other studies (Henderson et al. 2007: 92; Evans et al. 2004). In the previous chapter we saw the effect that unequal gender relationships or patriarchy had on the wellbeing of younger women. Here we can see an extreme example (but also an all too common one in the UK) of how such inequalities can take hold across a longer timeframe and its corrosive effects on one women's happiness.

Paula spoke repeatedly of the bad luck and misfortune she had experienced and how this had shaped her life and happiness that echoes the way that classical writers like Aristotle spoke of fate and Gods shaping happiness (Aristotle 2009). Her parents separated when she was a young child and with her two sisters and brother she went to live with her father. When she was eight years old he died suddenly and she was taken into care by the local authority, fostered with a family until she was 16 when she moved to her own flat. Paula spoke of her childhood as one marked by insecurity and feeling unloved by her foster parents; she seldom saw her mother and underachieved at school. She spoke of feeling angry and rebellious as a teenager.

> My foster parents already had two children of their own … Felt like an outsider in that family. One thing I hate is not being accepted. Feeling like an outsider I hated that. (Paula)

When charting these early events, there were some positive experiences—being helped to set up her own home in a small flat by her social worker and establishing a network of friends who shared an interest in music, dancing and clubbing. These early relationships came to be very important to Paula in later life—forging as a teenager some lifelong friendships that were a source of support that aided her wellbeing as an adult (see Pahl 2007). These difficult early life course transitions and problematic wellbeing reflect the poor resources and opportunities that many 'looked after' children experience (Barnardo's 2006; DfE 2014). Unlike the middle-class James and the working class Kevin, Paula didn't have the material, cultural or social capital that comes with a secure family background. Here we see the profound, cumulative effects of class disadvantage on the quality of life of young people in Britain (Roberts 2009). Her attainment at school suffered without the support of parents and then her efforts at finding work were hampered by her poor qualifications and the lack of contacts that might have eased her route into waged employment (see Webster et al. 2004). Consequently, Paula moved from one low-paid job to another during her teenage years, making do as best she could on her low income to pay for her rent, food and bills. At 18 years of age after a short relationship Paula became pregnant, but this relationship broke down and she lived as a single parent until meeting her long-term partner and later husband. When looking back over her life Paula sees the 28 years spent with her husband in mostly negative terms. The first few years of their relationship were happy but then her husband became increasingly controlling and then violent towards her.

> My marriage was just about being bashed around, it was total shit, total shit. I get angry now thinking about it. I must have been one idiot putting up with all that crap, Hey Ho … I took some beatings off the bastard, one thing he did in the garden, he threw all this water over me from the watering can. That hurt me more, I know it sounds silly but because it was in public—he humiliated me. (Paula)

As research into intimate partner violence has documented, violent relationships can be especially corrosive of women's wellbeing creating all manner of negative emotions like guilt and shame that compound the suffering (Golding 1999; Ellsberg et al. 2008). Paula often felt unsafe in her own home and her husband was emotionally abusive. These repeated acts where she was threatened and ridiculed undermined her self-worth so that she felt she was to blame for her husband's behaviour.

> I've put my hands through windows, got out of cars while he was still driving—just to get out of the car to get away from him. He was just in my head, 'wah, wah, wahwahwah'. Having a go all the time, moaning about anything and everything, I just had to get out of there … I would just sit in the park for hours … I'm just glad that I was a stronger person that I thought I was and what a lot of people might think… How some of this hasn't put me over the edge, I just don't know? … When you are living like that all you see forever, is this life, there's no escaping it … If it wasn't for my daughter and bringing her up it probably would have done. I would have just said that the world isn't worth living in and taken the easy option. I know I shouldn't be saying that but because I had her I didn't. (Paula)

Over many years Paula's wellbeing deteriorated markedly because of the abuse so that she considered suicide on several occasions. In her mid-twenties she was also diagnosed with multiple sclerosis. She suggested that on reflection there might have been some link between the anxiety and depression she experienced during this time and the episodes of illness that she had with the disease.

Surviving Abuse

Here we get a glimpse of one of the ways Paula managed to survive the years of abuse and maintain some sense of wellbeing—by investing herself in a range of roles, relationships and activities outside of her marriage. There are echoes here of the efforts that other interviewees made to live well by developing wider interests and social networks which helped counter some of the more problematic experiences in life (Layard 2005: 68; Pahl 2007). These also illustrate some of the creative ways that people can come to manage threats to their wellbeing as Goffman has shown (Goffman 1963). In particular, Paula spoke of the importance of raising her daughter, being a grandmother and the long-term friendships she had established when young. All of these activities contributed positively to Paula's wellbeing. Another way of coping was through small acts of defiance or resistance that allowed to her take some control back from her husband.

> His answer to everything was money … he would try and buy me. But I would never spend it on myself, I would loan it to other people instead … I hate money, doesn't mean a thing to me. I see it as being bought off … his way was like, give you money. I'll give you this to shut you up. I never kept any of it, always gave it away to my daughter. I didn't want it. (Paula)

The money symbolised the power of her husband and how he abused this in their relationship. By rejecting these gifts she had some small sense of control that she otherwise lacked. Significantly, this understanding of money has stayed with her and she still feels, despite changes to her personal life, and popular view on wellbeing, that affluence does not bring happiness.

Another way that Paula tried to cope with the years of abuse was through concealing the abuse she experienced—trying to maintain a sense of normality for her daughter, grandchildren, friends and neighbours. Only a few close friends were made aware of the violence that she dealt with during 20 years of marriage.

> My sister said recently, 'It's good to see you with a proper smile now rather than a fake smile, don't ever let me see you with a fake smile'. People did know, the ones that mattered but not everyone. But now I do smile for the right reasons, it's a genuine smile and I am genuinely happy … I used to smile sweetly because the kids were coming around … I wasn't true to myself. Haven't been true to myself for a long time. (Paula)

As we saw earlier with James, we also see here an example of happiness functioning performatively. Paula had learnt how to play the role of a happy housewife with a specific script, appearance and props (Goffman 1956). When her daughter and grandchildren visited she would smile and make conversation concealing how her husband had been violent to her just hours earlier. Given the shame that she felt as an abused women, one interpretation of the role playing here is Paula's efforts to preserve some dignity and respect in front of significant others—'saving face'. Although these performances—the wearing of masks to conceal the reality of her life—may have positive effects, at the same time as Paula recognised herself it also meant that she was alienated from her more authentic self. Maintaining such a secret from others must have taken some effort and also meant that these people who could have supported her were unable to at times when she surely needed it.

Paula would admit that her wellbeing overall had been poor, when compared to the likes of Kevin and James. The combination of misfortune, class disadvantages and years of abuse had undermined any chance she had of living a good life. This corresponds with much of the existing literature that links domestic violence and class disadvantage to poor wellbeing (Ellsberg et al. 2008). Yet, just as misfortune shaped her earlier life so good fortune also transformed her later life. Three years before I interviewed Paula she had met an old boyfriend she had lost touch with. Through social media she had managed to strike up a relationship that blossomed 30 years after they first became friends. Her new partner helped her to leave her husband and they have set up a new home together which offers her the security, love and happiness, that was missing from her earlier life.

> I had money, a daughter and grandkids but I wasn't happy, but I'm happy now. I wouldn't take this for granted as I know what it is to have a shit life. A lot of what went on then is still here [points to her chest, then her head]. It always reminds me that I didn't have my own life … I can go into town now and not have to worry about coming back and having to explain where I have been and who with. I have freedom that I didn't have before. (Paula)

Happiness and a Bohemian Way of Life

The other two interviewees in my mid lifers sample, Nigel and Kate, had sustained a good level of wellbeing for much of their adult lives—more so than the others in the mid-life sample or indeed younger samples. They scored very highly in many of the standard wellbeing questions (8–9 out of 10) and the qualitative data pointed to contented lives over the last 30 years or so. Yet when we examine their material situation and resources both appear to live very modestly. Both worked part-time in low-paid jobs, lived in small houses in disadvantaged communities, owned very old cars and spent little on household goods or lifestyle accessories like TV, furniture, mobile phones and holidays. They both seem to have pursued a happy life outside of the conventional Western notion that affluence and consumerism is the route to good wellbeing. How then have they managed to sustain this alternative route to wellbeing?

If we compare Nigel and Kate with the other mid lifers we can see some immediate differences in the respective lifestyles that might account for their good wellbeing. In comparison to the more conventional middle-class lifestyle pursued by James, both Kate and Nigel appear to have a far more balanced approach. Where James saw himself as the main breadwinner for the family and worked excessive hours to pay for an expensive lifestyle, Kate and Nigel consciously lived more modestly, working part-time in less stressful employment (Hamilton 2003). Nigel had worked as a bookseller, college tutor and pottery maker and Kate had worked as a teaching assistant in various local schools. They had come to work less in waged work than others as they had both struggled in the past in full-time jobs. One reason they perhaps avoided a mid-life crisis was that they had changed their lives for the better when young. Nigel discussed one of his first jobs as a teenager,

> So I thought I would be a psychologist … I got accepted at a psychiatric hospital, nursing and then almost immediately I was in conflict with the authorities because I had long hair … I ended up wearing it in a ponytail but got kicked out in the end, as I had a poster on my door, just so petty … They weren't as compassionate as I thought they should be … I would instigate little programmes and things to help bring people on and the rest of the staff would hate me for it, because it showed them in a bad light… the majority in these hospitals might as well be working for Ford … I was kind of persecuted all over the hospital really, they just seemed to want rid of me. But the revelation I had at that time … In psychiatry they say that you have to look inside people to see what is wrong with them, inside their heads. But they never did look the other way, at life, to see what life they were leading. I chatted to those people on the wards and I thought that 'God if I had a life like that I would have topped myself…' just aggression, mundane aggression and violence and the lack of any sort of compassion out there. (Nigel)

Nigel spoke of how he wanted to work with people and make a difference to their lives—hence his choice of nursing as a satisfying and rewarding job. This illustrates a key difference between Nigel (and also Kate) and the others in the mid-life sample and that is how they consciously tried to live according to particular values and ethical principles that reflected compassion, altruism, environmentalism and socialism. Interestingly, in his review of the literature Layard (2005: 71–73) notes how strong beliefs can enhance wellbeing. However, the bureaucracy and authority structures of many workplaces showed Nigel that work is often alienating for both patients and employees. This was a key moment for Nigel steering him away from the idea that a traditional career should be the primary basis for his identity and his wellbeing. We get a sense here of how slowly these experiences helped shape his values and ethics that have informed later choices about how to live well and work at happiness.

Another key reason why both Nigel and Kate had been more contented than James and Paula was that they both had sustained long-term relationships with their partners. When quizzed about this success they spoke about the importance of sharing interests and values with partners (see Waite et al. 2009). A keen interest in socialism and environmentalism underpinned their relationships and way of life. They lived modestly balancing waged work with wider interests and activities that supported other people—they were altruistic and compassionate. They also felt it was important that their children should share these values, were disciplined, respected others, were honest and had integrity. Nigel's happiness comes from how his son, partner and himself all share a similar set of values about how to live well—a collective, shared, social happiness.

> You hold your family together and the family holds you together as well … all three of us have a sort of shared outlook … so there has never had to be major compromises. (Nigel)

Both Nigel and Kate spoke of the difficulties of leading an alternative lifestyle—often other people were sceptical or mocking of a way of life that eschewed the conventional consumerist trappings of life. It takes a degree of discipline to live a thrifty life and a vigilance to monitor it to ensure it is balanced. Although Nigel was unaware of it this is an ethos that many classical writers on happiness would recognise from Plato, Aristotle and the Stoics through to later Roman and Christian theologians (McMahon

2006). They both spoke of having been helped to do this by the example and encouragement of their parents—key resources that were lacking in some of the other mid lifers that I interviewed. Nigel's parents, for example, were trade unionists and political activists themselves and Kate's mother (her father died when she was a small child) was a teacher and was also politically active. Both families valued education, debate and discussion and Nigel and Kate were fortunate to be given the space and freedom (and unconditional love) that allowed them to explore different ways of living. These are important sets of cultural and social resources, stemming from the upper working-class habitus that enabled both Kate and Nigel to explore different ways of getting on in life—opportunities that are often denied to many working class young people.

> It's also what goes on around you too—my dad was a trade unionist, a steward down the mines... They are a big part in keeping me together, my integrity and sense of self ... I remember my mum saying that I should get a trade and my dad saying, 'You shouldn't. I'm a time served painter and decorator and detested it, you should do something that you feel more comfortable in, you shouldn't saddle yourself up with something for life with one career. You should be more imaginative'. (Nigel)

When mapping out their social networks it was apparent that both Nigel and Kate were involved in a rich network of friends, interests and family connections that suggested a more balanced life than the career-focused Fifty Somethings like James. Much research around the world has linked work-life balance and greater control of daily schedules to better wellbeing (Goodin 2008; Thin 2012: 208)—something that was at the heart of Nigel's and Kate's way of life. Indeed Nigel and Kate also illustrate how their younger lives were very different to those Thirty Somethings we have seen in this research.

> I have a regular routine, Mondays I do this (coffee with friends), Tuesdays I do the choir, then there is the reading group I'm in—we meet in the church, I started that through my friend. And I'm in the Green Party, CND, and Amnesty International and also do things for the Humanist Society in the North East ... I jealously guard my own life—and I work my 30 hours. And I like it, it suits me, I go home when the children do. I arrive

> at 8.50 with the children and go home at 3.20. And it suits me and I have a life. (Kate)

Kate visited elderly neighbours each week, sang in a choir and had a circle of local friends who she visited every few days. She also took her 'thrifty' lifestyle very seriously, mending and making cloths and also knitting and sewing for pleasure. She had a den at the top of the house that was stuffed with her sewing gear and mementos she had collected over the years. She also spent much time scouring shops for food bargains and planning out different meals to use these ingredients—living well but cheaply was a challenge she enjoyed. Kate spoke of being a 'good manager' of the household budget and took great pride in running the home very frugally—something that was beyond most other people who lived more traditional, wasteful and 'throwaway' lifestyles. Even though she earned only £10,000 a year she was still able to save a few hundred pounds for a rainy day.

Nigel too was able to avoid full-time work by living carefully—spending much of his time repairing and selling on electronics, domestic appliances and cycles. He also repaired cars for people that generated some income. Both Nigel and Kate were also politically active in local campaigns for workers' rights, disarmament and environmentalism, developing a network of friends through these activities.

> You either earn the money to buy the things you feel you need, or as I said earlier that you reassess your needs—or you find different ways of getting them—recycling ... it's more re-use than recycle, I just bought a ghetto blaster for a tenner out of a second hand shop and its great ... with the HiFi, only cost me twenty quid, loads of bits from skips, charity shops and it sounded magnificent ... in part it was having time to go out and find these things in skips and shops. If you are working you don't have time to do that. (Nigel)

> I wouldn't want my happiness to be at the expense of anybody else ... I have always thought I would happily share, yes. I don't give it all away. I've built a home; I don't live an ascetic life ... I use secondhand things and charity shops and so on. I'm not materialistic in that way ... I think that is

> the secret of a good life, a happy life. I rather like the tagline—you know I said I was a humanist? The envelopes on their labels has the tagline, 'think for yourself, act for others' and I like that it's a nice little slogan. (Kate)

Over many years both Nigel and Kate had shaped a life for themselves that involved a shift away from working longer hours for ever greater income. Instead they had learnt to manage with less income, allowing them greater time for other activities that were important to them. In many ways they had embraced aspects of Eastern philosophy (such as Buddhism) as well as political and ethical principles rooted in socialism and environmentalism that suggest that a happier life is possible if one comes to manage one's desires and live a more balanced and modest existence (Schooch 2007). Like some of the young people I interviewed (Alice in Chapt. 7 who was a Christian) holding onto a set of values and principles for living offered a framework for making choices in life. Whereas, in contrast, James struggled to make sense of how to deal with life's challenges. His adherence to a more conventional set of values around the Protestant Ethic failed to offer the insights and support needed to deal with the complex problems he faced with his wellbeing. Hence, in recent years he had become interested in different values and some therapies like mindfulness techniques that offered different ways to understand the nature of his wellbeing (Choe et al. 2015).

A striking feature of both Kate and Nigel was how busy they were in a diverse range of activities. How they both embraced the opportunities that each day would offer them to achieve something new and different. I had a sense of this when I asked them about taking holidays; they both replied that they did not have holidays as there was so much to do each day. They were happy as they were so did not need holidays like other people. If we apply Marx's ideas on flourishing to Nigel and Kate we can see how their lives allowed them to develop their wellbeing in a variety of ways. They both had jobs that allowed them some control over how they worked and added value to other people that in turn rewarded them. Their interest in recycling and thrifty, sustainable living also allowed them some control of a productive process that created things of use. A large part of their daily routines were also comprised of relationships outside of waged work so they were connected to others at various levels—families, neighbourhood and region. Their hobbies and interests in local politics and environmentalism also allowed them to learn and develop new skills and some of these also connected them to the natural world around them. Taken together we can understand how their daily routines afforded them opportunities to experience many positive emotions, a sense of meaning and time to reflect positively on their lives.

Mid-lifers Looking to the Future: Being Fallible, 'Silver Linings' and 'Dark Clouds'?

The mid-life sample offered up some surprising insights into the happiness of people from different social backgrounds—counter to many of the arguments offered by other approaches to wellbeing/happiness. Despite his affluent lifestyle and success at work, James had spoken of the difficulties in his marriage and how they undermined his wellbeing. On reflection, he could also see that his long hours work culture and neglect of the rest of his life had also threatened his happiness during his forties and fifties. Compared to the rest of the mid lifers he was rather more socially isolated and had fewer friends than Nigel, Kate, Kevin and Paula (Pahl 2007). The other mid lifers who all led much more modest lives had been much happier. Even though Kevin had relatively meagre support from his parents when growing up this working-class upbringing did not undermine his ability to be happy as an adult. Though he had modest educational achievements which impacted on his job prospects, he did grow up within a complex network of family and friendship ties that supported his wellbeing throughout adulthood. Although social class is often viewed by sociologists as the cause of disadvantage in later life, here we can see some of the positive effects of class cultural traditions. In contrast Paula's more extreme experiences of class and patriarchal processes set up vulnerabilities early on in life that had a lasting effect on her wellbeing as an adult. In particular, her lack of a wider support network, transitions through education into poor quality employment and early motherhood all framed her route into an abusive relationship and very poor wellbeing. Yet despite the odds, good fortune had played a role in transforming her wellbeing in recent years. The happiest members of the sample were Kate and Nigel who had avoided a mid-life crisis by developing a bohemian way of life. Like Kevin, they benefited from a stable family background when growing up which aided their transitions into adulthood. In particular, they were shaped by sets of values and an ethical approach to living that guided them though life and supported their wellbeing. Despite their low income, the values they acquired from parents helped them establish the social networks and balanced life that was key to a sustainable quality of life.

A biographical approach can offer us an insight into the complex way that wellbeing emerges over time through the relationships people have with others. Although I have traced the ebb and flow of interviewees' wellbeing as they age, their stories reveal different tensions and currents that can shape their future happiness. James and Paula illustrate some of the inherent resilience that we all have—to come through challenging events and with time to begin to rebuild a life and one's wellbeing. James had cut his hours at work and this has allowed him time for new interests, hobbies and friends—cycling, gym, music. Though financially he is worse off than before and early retirement seems unlikely he has more time to lead a more balanced life than when he was married. Paula, too, after years of poor wellbeing feels optimistic about the future, despite her low income and occasional poor health with her multiple sclerosis. For Paula she feels free from the controlling behaviour of her former husband and this is key to understanding why she feels so happy despite her objectives circumstances.

Kevin too faced some threats to his wellbeing when one looked to the future. His beloved cycling club was changing—with newer and younger members joining and changing its activities. Kevin was upset that he was no longer central to the club and like many of the older members wished that he was ten years younger and faster. His aging also posed challenges for his work as gardening was physically demanding. He had begun to work fewer days as one way of coping with these changes. With the growing cost of living in recent years he was also economising—building his own little bar in the garden and inviting friends around for a beer rather than nights out in the town.

For Nigel and Kate they both hoped to continue their balancing of part-time work and the rest of their lives. Both spoke of the increasing pressure at work to comply with targets that undermined the pleasure they got from their jobs. Other concerns emerged from their families—Nigel's parents were increasingly frail and he spent more time caring for them than in the past. For Kate her daughter had moved to university and she was finding ways to manage this loss to her life—Facebook chats were one way of adapting to this.

Although those in mid life are all working in their different ways to live well at the same time their ability to adapt and thrive is constrained by the fallibility that is common to us all. The anxieties expressed by my interviewees about the future reflect some of this uncertainty they have about how well they know themselves and the way that their lives may change. My interviewees' stories reveal how we develop routines and habits of living that allow us to suspend our critical faculties and just get on with the day. Yet the familiarity of ordinary living as C. Wright Mills noted (1959) does blind us to small shifts in our world that over time come to influence us in significant ways. Commentators such as Kahneman (2011), Dolan (2015) and Deary (2014) have all documented this feature of living where we develop what seem innocuous routines that with time become cognitive blinkers that hinder all ability to make good choices and be happy. As a result this quest for happiness and a good life can seem like a confusing riddle as we are repeatedly surprised by how events turn out and how we feel about the consequences of these choices we have made. As these accounts illustrate there often appears to be unintended or unanticipated consequences of earlier choices and experiences that can be unsettling. Despite her new life, Paula spoke of her doubts about whether such happiness would continue—this unease reflecting the hidden legacy of her earlier life of abuse. At some level of her consciousness she was aware that she may struggle to erase the emotional scars of her earlier life and they may surface to threaten her newly found happiness. Kevin too—though speaking positively of his life—expressed doubts and anxieties about this continuing into the future. The dark clouds here were gathered around his relationships at the cycling club that had been evolving over many years, moving away from the one he had known and loved as a teenager and threatening his wellbeing. Yet fallibility can also bring pleasant surprises as James's story reveals—the silver lining of the turmoil of mid life has been the realisation that he has the opportunity to forge a new life for himself.

Bibliography

Adorno, T. (1941). On popular music. *Studies in Philosophy and Social Science, New York: Institute of Social Research, IX*, 17–48.

Aristotle. (2009). *Nicomachean ethics*. Oxford: Oxford University Press.

Bacon, E. (2015). *Great British cycling: The history of British bike racing*. London: Bantam Press.

Barnardo's. (2006). *Failed by the system: The views of young care leavers on their educational experiences*. London: Barnardo's. Retrieved from http://www.barnardos.org.uk/failed_by_the_system_report.pdf

Barter, D. (2013). *Obsessive compulsive cycling disorder*. Swindon: Phased Publications.

Bentham, J. (1776). *'Preface to the first edition', A fragment on government*. In Mack, M. P. (Ed.). (1969). *A Bentham reader*. New York: Pegasus Press, p. 45. In McMahon, D. (2006). *Happiness: A history*. New York: Grove Press, p. 212.

Blanchflower, D., & Oswald, A. (2008). Is wellbeing U shaped over the life cycle? *Social Science and Medicine, 66*, 1733–1749.

Bourdieu, P. (1977). *Outline of a Theory of Practice*. Cambridge: Cambridge University Press.

Bourdieu, P., & Passeron, J. C. (1990). *Reproduction in education, society and culture*. London: Sage Press.

Brown, P. (1987). *Schooling Ordinary Kids: Inequality, Unemployment and the New Vocationalism*. London: Tavistock.

Children's Society. (2015). *The good childhood report 2015*. London: Children's Society.

Csikszentmihalyi, M. (2002). *Flow: The classic work on how to achieve happiness*. London: Harper and Row.

Deary, V. (2014). *How to live*. London: Allen Lane.

Dolan, P. (2015). *Happiness by design: Finding pleasure and purpose in everyday life*. London: Penguin.

Dorling, D., Rigby, J., Wheeler, B., Ballas, D., Thomas, B., Fahmy, E., et al. (2007). *Poverty, wealth and place in Britain, 1968–2005*. Bristol: Policy Press.

Durkheim, E. (2014). *The division of labour in society*. New York: Free Press.
Easterlin, R. (2003). Explaining happiness. *Proceedings of the National Academy of Sciences, 100*, 11176–11183.
Ellsberg, M., Jansen, H., Heise, L., Watts, C., & Garcia-Moreno, C. (2008). Intimate partner violence and women's physical and mental health in the WHO multi-country study on women's health and domestic violence: An observational study. *The Lancet, 371*(9616), 1165–1172.
Evans, J., Rich, E., & Holroyd, R. (2004). Disordered eating and disordered schooling: What schools do to middle class girls. *British Journal of Sociology of Education, 25*(2), 123–142.
Freud, A. (1992). *The ego and the mechanisms of defence*. London: Karnac Books.
Gilbert, D. (2006). *Stumbling on happiness*. London: Harper Perennial.
Goffman, E. (1956). *The presentation of self in everyday life*. Edinburgh: Edinburgh University Press.
Goffman, E. (1963). *Stigma: Notes on the management of spoiled identity*. London: Simon and Schuster.
Golding, J. (1999). Intimate partner violence as a risk factor for mental disorders: A meta-analysis. *Journal of Family Violence, 14*(2), 99–132.
Goodin, R. (2008). *Discretionary time: A new measure of freedom*. Cambridge: Cambridge University Press.
Hagerty, M., & Veenhoven, R. (2003). Wealth and happiness revisited: Growing wealth of nations does go with greater happiness. *Social Indicators Research, 64*, 1–27.
Hamilton, C. (2003). *Downshifting in Britain: A sea change in the pursuit of happiness* (The Australia Institite, Garden Wing, University House, ANU ACT 0200, http://www.tai.org.au.)
Hatch, S., Huppert, F., Abbott, R., Croudace, T. Ploubidis, G., Wadsworth, M., et al. (2007). A life course approach to wellbeing. In J. Haworth & G. Hart (Eds.), *Wellbeing: Individual, community and social perspectives* (pp. 187–205). London: Palgrave.
Henderson, S., Holland, J., McGrellis, S., Sharpe, S., & Thompson, R. (2007). *Inventing adulthoods: A biographical approach to youth transitions*. London: Sage.
Hochschild, A. (2003). *The managed heart: The commercialisation of human feeling* (2nd ed.). Berkeley: University of California Press.
Hockey, J., & James, A. (2003). *Social identities across the life course*. London: Palgrave.
Jenkins, R. (1996). *Social identity*. London: Routldege.
Kahneman, D. (2011). *Thinking fast and slow*. London: Penguin Books.

Kahneman, D., & Deaton, A. (2010). High income improves evaluation of life but not emotional wellbeing. *Proceedings of the National Academy of Sciences, 107*(38), 16489–16493.

Layard, R. (2005). *Happiness: Lessons from a new science*. London: Penguin.

Layard, R., & Dunn, J. (2009). *A good childhood: Searching for values in a competitive age*. London: Penguin.

Levinson, D. (1991). *Season's of a man's life*. New York: Ballantine Books.

Lunau, T., Bambra, C., Eikemo, T., Van der Wel, K., & Dragano, N. (2014). A balancing act? Work-life balance, health and well-being in European welfare states. *European Journal of Public Health, 24*(3), 422–427.

Mac an Ghaill, M. (1996). What about the boys? Schooling class and masculinity. *Sociological Review, 44*(3), 381–397.

Mack, J., & Lansley, S. (2015). *Breadline Britain: The rise of mass poverty*. London: One World.

Marcuse, H. (2002). *One dimensional man: Studies in the ideology of advanced industrial society*. London: Routldege.

Mills, C. W. (1959). *The sociological imagination*. Oxford: Oxford University Press.

ONS. (2016, February). *Measuring national wellbeing: At what age is wellbeing highest?* London: Office for National Statistics.

Pahl, R. (2007). Friendship, trust and mutuality. In J. Haworth & G. Hart (Eds.), *Wellbeing: Individual, community and social perspectives* (pp. 256–270). London: Palgrave.

Roberts, K. (2009). Socio-economic reproduction. In A. Furlong (Ed.), *Youth and young adulthood: New perspectives and agendas* (pp. 14–21). London: Routledge.

Roberts, S. (Ed.) (2014). *Debating modern masculinities: Change, continuity and crisis*. London: Palgrave.

Roberts, S., Stafford, B. Duffy, D., Ross, J., & Unell, J. (2009). *Literature review of the impact of family breakdown on children*. European Commision.

Schooch, R. (2007). *The secrets of happiness: Three thousand years of searching for the good life*. London: Profile Books.

Seligman, M. (2002). *Authentic happiness: Using the new positive psychology to realise your potential for lasting fulfilment*. New York: Free Press.

Stevenson, B., & Wolfers, J. (2008). *Economic growth and subjective wellbeing: Re-assessing the Easterlin paradox* (NBER Working Paper No. 14282). National Bureau of Economic Research, MA, USA. Retrieved from http://www.nber.org/papers/w14282

Thin, N. (2012). *Social happiness: Theory into policy and practice*. Bristol: Policy Press.

Waite, L. J., Luo, Y., & Lewin, A. C. (2009). Marital happiness and marital stability: Consequences for psychological wellbeing. *Social Science Research, 38*(1), 201–212.

Webster, C., Simpson, D., MacDonald, R., Abbas, A., Cieslik, M., Shildrick, T., et al. (2004). *Poor transitions: Social exclusion and young adults*. Bristol: Policy Press.

Weis, L. (1991). *The working class without work*. London: Routledge.

Wilkinson, I. (2005). *Suffering: A sociological introduction*. Cambridge: Polity.

Wilkinson, R., & Pickett, K. (2009). *The spirit level: Why equality is better for everyone*. London: Penguin.

10

Happiness in Old Age

In this final empirical chapter I examine how four older people, in their sixties and seventies, experienced happiness and worked at living well. Research into aging documents the inevitable physical and psychological decline that we will all experience—impairments to our sight, hearing, mobility, a slowing metabolism and weight gain and decline in cognitive function. Later life also involves changes in status and social relationships such as retirement and bereavement that can shrink our networks and activities that influence wellbeing. Older people also have to manage an ageist culture in Western societies that involves pejorative constructions of older people as embarrassing and a burden on society. However, as I go on to show people can adapt differently to these challenges so that some age more successfully, living well and enjoying later life despite the inevitable restrictions of old age. Some of these differences can be traced to earlier life events. Hence, a biographical approach to wellbeing illustrates some of the cumulative effects of life experiences that shape happiness in old age. In particular we see some of the consequences of earlier class and gender processes for life course transitions and wellbeing. The women I interviewed had interrupted labour market careers and thus were less affluent than their male counterparts. The working-class respondents also

M. Cieslik, *The Happiness Riddle and the Quest for a Good Life*,
DOI 10.1057/978-1-137-31882-4_10

had accumulated less material resources than their middle-class peers that impacted on living standards and wellbeing in later life. Nevertheless, as we saw in earlier chapters, one could not always assume that affluence or success at work brings happiness. The older interviewees were no different in this respect illustrating some of the complex ways that happiness, resources and life events are interwoven.

Mary and Jill

Mary was a middle-class woman who was 90 years of age when interviewed, having been born in 1921 whilst Jill, born in 1934, was working class, aged 78 when interviewed in 2012. Mary had grown up in South Wales and London, her father had been a shopkeeper and she had worked as a seamstress in aircraft factories, later marrying in 1960 when she was 39. Her husband Derek was a manager in a company that manufactured retail signs and display units. He earned a good salary that enabled them to buy a house, take overseas holidays and allowed Mary to be a full-time housewife. They did not have any children. Derek died in 1983, and Mary then moved to South Wales to live near her two sisters where she spent the rest of her life living in a small but mostly affluent Welsh town. Mary died in January 2016 at the age of 95.

Jill worked as a shop assistant when a teenager, and her interest in first aid (with the charity, the St John's Ambulance) led to jobs as a nurse and first aider. She was married as a teenager, had one son and worked with her husband managing local pubs and clubs in South Wales during the 1960s and 1970s. Her husband died suddenly at the age of 50; she then worked as a cook and manager in pubs before retiring. Jill spent most of her life living in South Wales, renowned for its docks, railways and steel industry—most of which have now disappeared. She lives in one of the many working-class estates in a town that have suffered from high levels of unemployment and poverty since the 1980s.[1]

[1] The author conducted doctoral research in this locale in South Wales that examined the schooling experiences of young people from disadvantaged backgrounds—see Cieslik (1997).

Happiness and Wellbeing in Recent Years

I interviewed Mary and Jill during 2011 and 2012, and as they were both over 75 years of age I did expect them to have experienced some deterioration in their wellbeing. Research by gerontologists (Vincent 2003: 20–21; Bowling and Dieppe 2005; Green 2010: 175) suggest that the earlier part of old age, between 60 and 70, can be some of the best years of a person's life but that wellbeing declines after this point. This is confirmed by wellbeing surveys that suggest that many people's wellbeing improves after mid life, only dipping again as people approach their eighth decade (Blanchflower and Oswald 2008). This eventual decline in wellbeing can be attributed to the greater incidence of ill health and bereavement (such as the loss of a partner). However, there were marked differences in the respective wellbeing of interviewees—although Mary's wellbeing was one that might be expected at this stage in life, Jill by contrast said she still had a very full and rewarding life. A further surprise was that it was Jill who was living a very modest working-class life, whilst it was Mary who was the more affluent and middle class.

Both women completed a wellbeing questionnaire that showed some differences in their views on the quality of their lives (see Hoggard 2005). Jill said that she was happy 80 % of the time, whereas Mary suggested that she was happy only 40 % of the time. When asked how she rated different aspects of her life (on an 11-point scale) such as her health, family, home and leisure, Jill scored 8 or 9 out of 10 for many of these domains, whereas Mary tended to score these much lower, around 5 or 6 out of 10. When completing questions that measured their feelings or emotions towards aspects of their lives, once again Jill answered much more positively than Mary, suggesting that Jill was much more satisfied with her life than Mary. Jill's responses suggested that she felt much more in control of her life, felt valued by others and that her life had meaning and significance for her. Mary's responses were all rather more pessimistic indicating that she found less meaning in life, was less optimistic about the future, felt less energetic, less inspired and more anxious than Jill.

How can we account for these significant differences in the wellbeing of these two women? Why might affluence be no guarantee of happiness in later life? The mapping exercise I undertook charted the social networks and activities of both women and comparisons of these revealed significant differences between them, indicating why their wellbeing was so different. Jill had much more extensive social networks and activities than Mary whose routines involved just a few social engagements such as coffee with neighbours, shopping trips, and occasional visits from family members. Mary then spent a large part of each day at home, watching TV or listening to the radio. Although Jill had a son and two grandchildren who visited each week it was not her immediate family that shaped her social life but a bewildering array of activities, friends and networks. This meant that, unlike Mary, she spent most of each day away from her home, 'keeping busy' as she said. Jill said that she watched little TV, she joked about how she was busier and worked harder now than when she was at work. This corresponds with research that suggests that many women, unlike their male counterparts, do not retire when they finish work as caring commitments continue (Bernard et al. 2000: 30–31). Jill was the caretaker for the local community centre so organised the room bookings for different clubs and social events. She ran the weekly bingo at the centre as well as an evening dinner event. There were also regular charity activities and Jill usually baked cakes for these—she was well known in the area for her baking. Jill had also become an informal carer for some of her neighbours who were ill or recovering from bereavement and this involved cooking, helping with pension and benefits issues and running them around in her car. Jill also met up with friends for a weekly dance in the town centre that involved lunch and bingo. These women also organised regular coach holidays to seaside resorts—the so-called, 'Turkey and Tinsel' holidays in the run up to Christmas every year. When I asked Jill about an average day in her life she said,

> Yesterday, we had some druggies around and they [neighbours] were afraid to tell the police so I phoned them and told them my name, I don't care. Then someone else came over to me as she had trouble with a girl that she couldn't sort out. So I went over there to sort that out and then somebody came over to ask me to cook something for them. Some of the old ladies [at

> the community centre] then asked for some pasties and then some Welsh cakes. I just do it because if it's helping them … if it's helping them then I am happy. Because I don't go out, I don't drink, I don't smoke, I eat too many chocolates. Other than that I am just doing things for other people. I just don't think of it, I just do it … I always think that you get more back in the long run that you give, I do. (Jill)

Jill illustrates the importance of friends and altruistic or reciprocal relationships for good wellbeing. Aristotle spoke of the significance of friendship for a good life and how virtue is its own reward—that it may seem an effort to be kind and support others but such kindnesses are often returned, enriching one's life in the process (Aristotle 2009). This feature of Jill's life illustrates one of the themes of this book, namely the social nature of happiness—how we often spend our time working for the happiness of other people that can 'magically' aid our own wellbeing. Jill's community activities meant that she was well known and admired by neighbours which offered her a status on the estate where she lived making up for any loss of social position that occurred when she retired from waged work. This effort to continue with social activities that offer status and social standing has been noted by gerontologists as one of the hallmarks of 'successful aging' (Bowling and Dieppe 2005). This contrasts with earlier conceptions of old age which assumed some form of 'natural' and progressive disengagement from wider society (Green 2010: 178). Our conversation took in the difficulties of being old, having to accept that one was slower doing things than before and that loved ones were no longer alive. When I quizzed Jill about the importance of 'keeping busy', she inferred that it meant that she had little time to dwell on negative thoughts—the reality of aging, losing her husband, brothers, sisters and friends.

Keeping busy and active also meant Jill continued to use a range of skills—cooking, driving, dancing and running events such as the weekly bingo. One can see how these activities underpinned her flourishing—the baking, for example, which she did for several hours each day had all the hallmarks of 'flow' experiences (Csikszentmihalyi 2002)—concentration, challenge and also learning new skills as she experimented with different recipes. Researchers have also studied the effects of dancing on older people showing the range of benefits for psychological and physical wellbeing (Cooper and Thomas 2002). If we examine Jill's life in relation to Marx's ideas on flourishing we can see that the four dimensions he explores all feature in Jill's way of life. She employs a range of skills in a productive way, has some control over this process, the things that she makes have use value for her and the activities are also embedded in wider, beneficial social relationships across various domains.

In contrast to Jill we can see how Mary's more solitary life may account for her poorer wellbeing. Our conversations suggested she had more time to dwell on the difficulties of aging—the loss of her husband and siblings and how she rarely saw her nephews and nieces. She spoke to me about her interest in music, film and some TV programmes but the scope of her interests and their positive effect on her wellbeing were less significant than those we saw with Jill. Although she did not admit to being lonely her responses to the survey questions and the interview data all pointed to a desire to have more social contacts and opportunities to socialise with others. The sorts of experiences we see with Mary were documented by the Economic and Social Research Programme into Aging (Scambler et al. 2002) that suggested a significant increase in the numbers of older people in the UK reporting being lonely. These researchers have suggested that loneliness can hasten the onset of cognitive decline as well as poorer mental health amongst older people. Mary spoke of her experiences of old age,

> I moved here as my sisters lived across the way there. But Laura and Julia have both passed on now, all of my family, I'm the only one left now … I like sewing but I can't see to sew anymore. And I clean and I cook and that's about it … I get on with most people, but bad things keep on popping up all the time … I've had my life, we are like flowers, you are born, you grow, then you see the flower gradually, only gradually, if you watch a flower dying. And that's us, I mean, I look at my hands and I and I go 'Ugh—they're old'. And that's you growing old … I've lost all of my loved ones haven't I? I haven't got any of them left, they are all gone. (Mary)

As well as this rather more pessimistic outlook on life, the other significant difference between Mary and Jill was that Mary had fewer interests than Jill. Hence she has much less opportunity to use and develop skills, to be cognitively challenged in these activities, to create something of use and to meet people through these productive activities. When people spend large amounts of time at home they tend to then often watch TV for long periods too. Research on the effects of TV on wellbeing notes it encourages a passivity that can be corrosive of wellbeing and this was (Layard 2005; 86; Csikszentmihalyi and Leferve 1989) something that was a feature of Mary's routines.

A noticeable feature of the differences between Jill and Mary's life was how Mary looked forward to her weekly shopping trips with her neighbour, where they would buy provisions for the week and then visit shops buying clothes and expensive china. Mary's house, in a select retirement community was characterised by its neat and tidy arrangement of expensive furniture and the many china ornaments on tables and stands. Whereas Jill lived in a disadvantaged community and her home was more spartan with a simple old sofa and chairs with few ornaments. What she lacked in porcelain, however, she made up for in the photographs around the house that featured the many people she had known over the years. Her kitchen too was like a mini baking production line with cake mixes, bowls, trays of cakes and various cooking utensils.

Biography and Wellbeing

One of the aims of this book has been to show how happiness is something that people do through their lives with other people. Hence it can be structured by life events (that reflect shifting economics and cultures) and emerges over the biographies of individuals involving complex layers of meaning. So when people are asked about what happiness is or what a good or successful life is—their answers reflect these personal histories. Yet much contemporary research in this area employs simple questions and measurement of wellbeing that views happiness as something that you *have* rather than as a *social practice*. And so the complexity, the processes and meanings that give happiness its nature are not always captured by these efforts to measure it.

In the case of Mary and Jill, the survey questions I asked them did show us there were significant differences in how they rated their wellbeing.

But what the questions struggle to do was offer us a deeper understanding of their (un)happiness or suggest how these differences may have arisen. Hence there is a need to combine these sorts of survey approaches with further qualitative and biographical investigation. The qualitative data has suggested so far that the differences in the women's wellbeing might be linked to the differences in the social networks and activities of Mary and Jill. The two women also differ in that Mary is several years older than Jill and had experienced poorer health and more bereavement that could account for Mary's poorer subjective wellbeing. However, the biographical mapping exercise that I did with both women revealed how they experienced rather different life events that have influenced their wellbeing over the longer term. It seems that Mary's poor wellbeing may not just be associated with older age but a more general feature of her life when she was young, despite the material privileges of a middle-class upbringing. Whereas for Jill, despite her modest working-class background she was fortunate to have grown up with a range of cultural resources that encouraged a better subjective wellbeing. Although these life histories are unique, unrepresentative case studies, they do offer us insights into the nature of happiness and some lessons on how to live well.

When I questioned the two women about their early life they offered quite different accounts of growing up during the 1930s and 1940s. Mary's father was from a family of small business owners and the family ran a cobblers shop. She had aunts and uncles who were well-educated, prosperous and lived in large houses around Southern England. Mary said she grew up to be very conscious of the differences in social status between the working classes who lived at one end of the street and the professional classes of doctors and dentists who lived in the big houses at the other. She spoke of her family being, 'stuck in the middle' able to afford a nicer rented home than the rougher working classes but not as respectable as the professional classes. This status consciousness and insecurity about her social standing have stayed with Mary throughout her life. It was a characteristic that was reinforced by later events and which have had a bearing on her sense of self-worth and her wellbeing. This way in which her identity was shaped also involved her experiences of education. She had aspiring parents who expected their children to do well at school and become teachers, doctors or engineers.

Some of Mary's siblings did well, attended grammar school, university and entered the professions. Mary, however, had a much more difficult time at school.

> I just couldn't see the words … I wasn't a scholar I had a learning difficulty but so far as the teacher was concerned I was just not very bright at all … You're listening and you look like you are keeping up but the point was I didn't have the memory to keep it there … Mrs Davies treated me dreadful, she used to bring me out in front of the school and I remember a little boy sitting in the front and I started to cry and I had to show my paper and it wasn't very good and he said to me 'don't cry Mary, don't cry Mary' … She humiliated me by sending me to a lower class to show my book. Now how can a teacher get to do that to a child… she made me feel that I was no good and that I didn't try but I did try … I could never be anybody special because I hadn't the brains for it … I did object to it and I hated my teachers. (Mary)

To us today it would be apparent that Mary had dyslexia so that although she could read she struggled with writing skills and had cognitive difficulties that impaired her memory. In the 1930s, however, teachers were unaware of dyslexia so Mary was labelled by teachers sometimes as unintelligent and other times as lazy. Given her social background and high expectations of her parents these judgements by teachers left a deep and indelible mark on her self-esteem and identity—she was shamed and stigmatised which had an enduring legacy (see Cieslik and Simpson 2015).

Jill in contrast had a much smoother journey through childhood, growing up in a working-class family with parents employed in local industries. She was expected to attend school and study enough to enable the transition into traditional feminine jobs such as shop and factory work. After a much more positive experience of school than Mary, Jill gained some school certificates and took up her first job at 14 years of age as a shop assistant. Though they lived in cramped conditions on what to Mary would have seemed a rough working-class estate, Jill had an active role in the local church helping her mother organise community events. As the eldest of eight children Jill was expected to help out around the home and this resourcefulness was also employed in the wider neighbourhood to support the elderly, children and the sick.

> I was the eldest of eight children … I was left more of less to see to the others and I was always helping them out. In school I was always running errands for the teacher and cooking for the morning break, I've always been on the go. And I use to do other people's shopping … I think [the church] has helped me a lot because I went to church from the time I was little. I mean we had to go then because our mam made sure we went. But, as I grew up, I still went and I feel it gave me strength to keep going. (Jill)

Jill spoke of her close relationship with her mother and how she was a pillar of the local church and community—an example of how to live well. The good relationship with her mother, her emerging role as a community activist and the role of religious beliefs were all significant for helping Jill to develop a far more positive sense of herself than we saw with Mary. There are echoes here of Max Weber's analysis of beliefs and the role of meaning for good wellbeing—how religious values can help with the challenges of everyday life (Weber 1990). We have a view of religion here as significant for the wellbeing of the working classes—providing both moral and practical support for those who were struggling, like Jill, with large families, little money and poorly paid jobs. Of course many Marxists, as we saw in earlier chapters, would be sceptical of such a positive interpretation of the church in the life of the working classes. Instead the Marxist line suggests the wellbeing of people like Jill was superficial, obscuring the underlying structures and the power relationships in employment and politics that impoverishes the working classes (Marx 1984). How can we arbitrate between such divergent understandings of wellbeing and religion? At an interpretative and micro level individuals can offer positive accounts of religion and wellbeing as we have seen in interview data here, yet at a macro and more theoretical level commentators have developed more pessimistic interpretations of how wellbeing and religion operates. Do we conclude there are different ways of understanding how wellbeing and happiness functions—an aspect of the numerous paradoxes or riddles of happiness?

If we apply other traditional sociological theories to the interview data we see further curious accounts of the functioning of happiness. In Jill's case her working-class culture—her distinctive habitus, though restricting her educational experiences, as Bourdieu argues (Bourdieu and Passeron 1990)—did function positively in regards to her wellbeing. Whereas, the middle-class habitus and the ambitions for educational success and social advancement meant that Mary was always likely to be a disappointment. These observations about cultural background and the shaping of expectations illustrate some of the many conundrums of happiness studies. How someone like Jill can be raised to have modest expectations that might be easier to achieve and lead to happiness or like Mary be socialised into more ambitious goals which makes one vulnerable to disappointment and eventual unhappiness. Philosophers have long pondered such dilemmas around the nature of happiness, success and a good life—whether we can compare the happiness of a simpler person to that of a cultured one? Surely someone who has lived a long and varied life will experience a richer more valuable happiness than someone who has only known simple hedonism (Layard 2005: 22–23)? Once again, is it possible to answer such questions and develop some rules on how to live well as philosophers have tried? Or do we, as my interviewees have shown, rather acknowledge that happiness is not an abstract thing but something that we just keep on doing everyday—that we muddle on through as best we can?

Marriage, Love and Happiness?

Mary's adolescence was interrupted by the Second World War and she enlisted in the Women's Auxilliary Air Force, helping to repair airplanes. This she said was an enjoyable time, living away from home with other women and learning to develop skills in mechanics and sewing. At the end of the war she returned home and worked in a series of jobs that drew on her wartime experiences and skills. She also helped her father, who by this time was running a small corner shop. Though she had some romantic attachments during her twenties and thirties she found herself still living at home whilst her younger brothers and sisters had all married and moved out of the family residence. Eventually, she met her 'husband-to-be' at the age of 37 and married in 1960 when she was 39—her husband was 40 years of age. Looking back at this time I was expecting Mary to talk fondly of her husband and married life but instead, like her earlier educational experiences, it seems to have been one of regrets and disappointments.

> My mother always used to say, you might think you are in love but it's the chemistry between two people … She said that if you have never had it you don't know. You can see a person a long way off but you would know if the chemistry is there … I said to her what do you mean by that chemistry and she said that one day you will find out, well I didn't find out. I married Derek because I was 39, he was nice, he came from a nice family, he had a nice mother and father and we got on quite well … but there was no chemistry there. (Mary)

Rereading the interview transcripts it seemed that her life at the time appeared to be a successful one. Her husband earned a good salary in a respectable career that enabled them to purchase homes that grew in size over the years. They drove expensive cars and enjoyed holidays overseas at a time when few could afford such things. She could afford to leave work and be a full-time housewife. Yet now, looking back on this time, this exterior success and happiness seems to have been a façade. The material comforts of a middle-class life that she had aspired to did not make up for the fact that her marriage was not a loving one. There was something missing from their relationship—the deeper, authentic understanding and intimacy that comes with genuine love (Thin 2012: 119). Mary spoke to me a few months before she died and said that one of her lasting regrets in life was her loveless marriage.

Jill in comparison was fortunate in finding a partner to love and they were married and had one child. Though their relationship was cut short, like Mary's through the premature death of her husband, unlike Mary, however, Jill had the support of her mother, friends and family and was able to cope rather better than Mary with the loss of her husband. She continued to work in various jobs in the pub/club industry and also continued her community work on the estate where she had grown up. Mary in contrast, however, spoke of a lengthy period of depression after the death of her husband. She had not worked for many years and lacked the friends that waged work often brings. They had also moved house several times and so she lacked the networks of neighbours which had aided Jill's recovery from bereavement. As she felt little for the community where she had lived Mary moved in 1985 to South Wales to live near her sisters where she remained until her death in 2016.

Earlier Life Experiences and Successful Aging

By exploring Marys and Jill's life stories with biographical methods we can develop a more nuanced understanding of their wellbeing in old age. Aristotle wrote how one can only assess ones happiness towards the end of one's life—reflecting back on ones triumphs and disappointments (Aristotle 2009). We can appreciate the differences in the subjective well-being of these two women when we compare how events and experiences accumulate over time, shaping identities, life chances and happiness. Mary had a difficult time at school that undermined her self-confidence and deprived her of the qualifications that might have allowed her to have a career and independent life. Wishing to please her status-conscious parents, she married a respectable middle-class man and fulfilled her role as a housewife. A feminist reading of her life might suggest that Mary had little choice but to settle into the respectable housewife role as the patriarchal norms at the time offered few alternatives (Oakley 1974). As Sara Ahmed has argued, notions of happiness in Western societies are associated with certain normative ways of living—heterosexual love, marriage and 'domestic bliss' (Ahmed 2010). Although Ahmed's account tends to depict women as hapless victims of oppression, we do instead see from the interview data that Mary developed ways of coping with some of the disappointments in life. Her wellbeing was not as bleak as it appeared as she enjoyed the materialist trappings of middle-class life—the entertaining, the cars, holidays and luxury goods. The interview transcripts hint at a degree of impression management—Mary developing a personae or mask that she used to successfully perform this housewife role (Goffman 1956; Hochschild 2003). We can only speculate about the gulf between the inner world and desires that she held and this exterior self that she projected. With time, however, we can see how Mary came to realise that these performances did not bring the happiness she hoped for but instead a lasting regret. But this was an aspect of her life that she had hidden away and was reluctant to explore during interviews save for some cryptic remarks.

> [With my husband] the chemistry wasn't there. If the chemistry is there you will die for that person. What if that chemistry comes to you when you are older? Which came to me … This chemistry is a thing—I can explain it to you but I am not going to because that is something private … Yes I did get the chemistry, but not through Derek. I'll not tell that because it's a private thing for myself. (Mary)

Jill's transition to old age in many respects has been more successful than Mary's—she has been happier as her earliest experiences were shaped by her mother who offered her an example of how to be a strong, independent, working-class woman. This was a template for living that she carried with her throughout her life. Contrary to this notion that many women are oppressed as they have been seduced by the idea that happiness comes through marriage (as discussed by Ahmed 2010), we have instead a much more creative, activist model of femininity. Jill, like her mother before her, saw her domestic role as only one part of her life and so her better wellbeing emerged through her wider engagement with life—her different jobs, interests and community work inspired by her religious beliefs and pride in her neighbourhood.

Frank and Martin

In July 2010, I suffered a cerebral haemorrhage that meant I was unable to work for a year while I recovered from the physical and cognitive impairment that comes with brain injury. I was more fortunate than most as I had experienced just a minor bleed into the brain and so by early 2011 the headaches had subsided and I was able to walk again. As part of my convalescence doctors suggested I take up some art classes and so I enrolled onto a beginner's pottery class where I met Frank and Martin. Frank was 67 years old when interviewed in 2011, having been born in Nigeria in 1944. He had retired two years previously from his job as a self-employed builder and was living in a small rented property with his 12 year old daughter from his second marriage. Martin, who was born in 1940, was 71 years old when interviewed in 2011. He lived in a large terraced house with his wife in an affluent town in the North East of England.

When I examined the wellbeing survey questions that both men completed it was apparent that both scored themselves very highly for overall satisfaction with life. Martin said that he was happy 70 % of the time and Frank said that he was happy 95 % of the time. When responding to statements about their lives both men were very positive about their day-to-day wellbeing such as having meaning in life, feeling grateful, being able to change their lives for the better, having energy, having things to look forward to and so on. They both scored themselves highly (8/9 out of 10) for optimism, relaxed, joyful and suggested they rarely felt lonely, bored or depressed.

The wellbeing data suggests that Martin and Frank are as happy and content with life as Jill who we met earlier on in the chapter. This is interesting as it runs counter to the arguments of gerontologists (Van den Hoonaard 2007) that some older men have poor wellbeing as a lifetime of work has meant they have neglected the social networks and activities needed for successful aging. Other research (Blane et al. 2007) also suggests that material disadvantage working through class inequalities can also be a barrier to good wellbeing for older men. A lifetime of manual labour, smoking, alcohol and poorer diets can lead to poor health and unhappiness in old age. Yet Frank who arrived in the UK in late 1960s and worked in manual jobs for much of his life continues to have good wellbeing despite these sorts of experiences associated with a working-class background. One factor that may account for their relatively good wellbeing is that they are both young compared to many retired individuals in Britain. Unlike Mary, they are still in good health and have yet to experience the loss of loved ones and the grief that can often undermine wellbeing.

Many researchers have noted (Layard 2005; Pahl 2007) that having a structure of sociable, daily activities is significant for maintaining good physical and psychological wellbeing. These were features that I discovered when mapping out the social networks of Frank and Martin. Though there were challenges that the men faced pursuing a good life, Martin discussed his struggle to keep fit and healthy, as he had gained weight in recent years that made walking and gym work more difficult.

He also acknowledged the central role that his wife played in his life and how he was fearful of her becoming ill and the impact this might have on his life. Frank too had some health concerns as he had been diagnosed with diabetes and had to carefully monitor his diet. Frank said he had experienced some 'low points' in recent years as he had separated from his second wife and taken custody of his daughter. The separation had involved hearings with social services and a sharing of assets with his former wife—all of which had been very stressful. Though materially he was much worse off than previously (e.g. his former wife receiving the family home) Frank said that, three years after the separation, he was now much happier than at the time of the separation. This corresponds with wellbeing research that suggests that many people can surprisingly bounce back a number of years after separation and divorce (Layard 2005: 66).

Hobbies, Interests and Happiness in Old Age

The mapping exercise I undertook with both men illustrated the complex nature of their social networks and activities that, like Jill, may account for their relatively good wellbeing in old age. Both men regularly spent time with their children and grandchildren, took regular exercise, had a range of interests and hobbies and extensive friendship networks. They have the sorts of lifestyles that researchers suggest is associated with more successful aging (Bowling and Dieppe 2005; Ward et al. 2012). The two interviews with Frank were conducted at the large kitchen table where he entertained his friends to the various dishes he cooked. Where we sat, we were surrounded by the pots, pans and utensils he used in his kitchen. His approach to food is different to many today who view it as 'refuelling'. Instead Frank spoke about the complex associations that food and cooking had for him and its importance for a good life.

> I like to cook a variety of food, I do Chinese, I do Indian, I do African … Back in Africa all the boys have got to know how to cook, you've got to look after yourself … My daughter helps me when she comes from school and wants to make something. We do it together. It takes longer, but it's more pleasurable for her … So we love food, my whole family. My son cooks good meals. And now at home he is the one that does most of the

> cooking ... It's sociable, my sister was here last week, that's why they are all staying here now. I invited my friends and everything ... and it cements a relationship. If they come around, often, two or three times a year, we have the same people around again, and I go to their houses as well. (Frank).

Food and cooking was nourishing at a number of levels for Frank as the recipes were part of the traditions and rituals of growing up in Nigeria. An interest in food and the skills of cooking had become a positive part of his self-identity. He wanted his own children to grow up with an appreciation of food—the kitchen then was an important place for informal learning in the home. Frank said that cooking was a great way to spend time with his teenage daughter at a time when many 12-year-old girls might have tired of spending time with a 67-year-old man. The rituals and culture around food he wanted to pass on to his own children so that they learnt something about Nigeria and acquired important life skills. In the process he hoped they would come to see how important cooking and sharing meals are for a good life. The love of food and entertaining as we saw with Jill earlier said much about the values and ethos that framed Frank's life. He was generous—sharing meals with his friends and was also busy learning and developing his skills. The self-absorption and challenge of cooking again has the hallmarks of Csikszentmihalyi's flow activity that can be rewarding, enjoyable and enhance psychological wellbeing (Csikszentmihalyi 2002).

Martin, in contrast, spoke at length about his lifelong interest in art and how painting, drawing and sculpture were important for his wellbeing. He left his full-time job as a teacher when he was 55, and although he worked part-time since then he had also developed this interest in art. He had taken a part-time BA and MA in Art and also attended many short courses on different aspects of painting, pottery and sculpture. Like Frank and his love of cooking, Martin's home also reflected his interests but instead of a busy kitchen, Martin showed me his workshop full of canvases, paints and ceramics. As Martin had worked as a therapist for many years he had come to realise how expression through media can reflect tensions between hidden internal worlds and visible external personae. As we explored some of the images he had recently painted he offered some interpretations of these and what they might say about him and his wellbeing.

> I've always been creative, I've drawn from being a kid, I've always been drawing or painting or making sculptures ... This piece, it's called 'Tell me grandad', it's, I suppose my own relationship with my grandson but aesthetically it's got all the curves and flows but it's a little boy and a granddad and they have been there, done that ... It's like walking into a garden and there's a labyrinth around it ... It's just a representation of an internal way of being if you like ... it's symbolic of family life as well. I've enjoyed it, it's like getting, it's externalising an internal feeling. (Martin)

At a superficial level, Martin simply took pleasure in the act of drawing, creating paintings or moulding material into different shapes. He could lose himself for hours in his workshop and enjoyed this self-absorption that took him away from the usual daily thoughts and tasks. Over the years, he also had improved his technical skills, which was satisfying. He had made many new friends through the workshops and classes he had attended over the years, enjoying the socialising and discussion of art. As research has shown, when people retire they have to develop new interests to create the networks and friendships that will support their wellbeing (Scambler et al. 2002). If they struggle to do this, as we witnessed with Mary, then loneliness can be damaging to their physical and psychological wellbeing (Lawton et al. 1999; ONS 2015b). At a deeper level, his artwork was also nourishing as it allowed for free association—the ability to access some aspects of his inner self that are usually hidden from the conscious self. He felt such a process was cathartic, allowing him to develop a better understanding of his self through the expression of his fears and desires. The image of him and his grandson illustrated some of these ideas—how he now sees himself as an old man beside the young child—fearful of aging, ever conscious of lost youth and desiring more time with loved ones.

Frank and Martin: Situating Current Wellbeing in Their Biographies

One of the challenges for happiness researchers is grappling with how wellbeing is a complex set of experiences influenced by the interpretations and various meanings that actors construct in everyday life. Numerical representations of wellbeing popularised by economists and psychologists can only offer us a rather crude insight into how and why certain experiences or events are good or bad for wellbeing. We do need some context and history to make sense of how some people are happier than others and how those others might change their lives to live better. This is why self-help guides often fail to help us to live better as the general rules for living that they offer us lack the context and concrete detail we need to apply them to our own lives. Similarly, much wellbeing research founded on large survey data ends up producing such general insights (such as marriage and good quality work are good for wellbeing) that the findings appear crass or common sense. With Michael and Frank, we see the need for research that explores the layers of meaning that are significant for wellbeing. The way they lived their lives as older men and how hobbies and interests had particular poignancy for them had emerged out of the ebb and flow of their own personal histories. Survey research with its focus on measurement and correlations would have struggled to make these links between their biographies and their current routines and wellbeing. A biographical approach illustrates how apparently simple and innocuous interests in food or art have a deeper significance for them and their happiness.

Philosophers have explored the idea of 'the sublime' to denote how some experiences can have an extraordinary or intense emotional significance for individuals (Burke 2015; Kant 1951, 2003; Schiller 2004). Although these writers often spoke about the wonder that we feel at great works of art or the power of nature I suggest that in our everyday lives we can often experience this transcendental quality of the sublime when our personal histories imbue something with complex layers of meaning. We can have intense emotional reactions as we are connected in some way, because of our biographies, to places and people and our imaginations and memories confer this depth of feeling. As the cover of this book illustrates I have had moments on a beach that transforms these places into magical sites with enduring appeal. To revisit them again or relive it imaginatively helps us to travel out of the 'here and now' to somewhere else. To have these moments of the sublime are life-affirming, exercising our uniquely human skills of reflection, reminding us of what is worthwhile in life. Although for Frank and Martin their hobbies and interests were often experienced in routine, humdrum ways, they also spoke about them in ways that suggested that art and food had a powerful resonance for them—as sublime moments—which accounted for their nourishing, fulfilling or restorative role they had in their lives. To appreciate how these aspects of their lives came to have such significance for them and their wellbeing, we need to explore their life histories in more detail.

Although the detail of Michael and Martin's lives differs greatly, the structure to their biographies do share some common features. They both experienced sadness and disappointments and their efforts to make sense of some of these failures and learn from them accounts in part for their wellbeing in later life. They both grew up in the 1940s, socialised into traditional gender roles that assumed they would work hard, become a 'breadwinner' and support a wife and family. Yet Frank's migration to England from Nigeria, his experiences of racism and difficulty securing work posed real challenges for these expectations he had of himself and of a good life. Similarly, Michael faced structural barriers to his desires to get on in life. Though he was from a middle-class background he had been educated in a harsh Catholic school in Ireland and failed to win a scholarship at the preferred secondary school. He always felt that he underachieved and disappointed his aspiring parents (as we saw earlier with Mary), deciding instead of university to join the Royal Air Force.

> I think I always expected too much of myself so, you know, I was very self critical and that was founded I think way back, at home … I was always saying, Oh I can't do this, I had to do this, I need to get this done and I get to clear my mind and then we can focus on something else, but then in the meantime something else work orientated would come up … When I started off [working] in special needs I was very fit and I used to meditate twice a day that gave me balance … Sometimes I stayed later at work so that got lost and this slowly built up over a number of years. I was teaching in that environment for 14 years which is far too long for anyone … I could see the children being wasted … and the frustration of not being able to do something about it. (Martin)

> It was every day, my friend, every day. Black bastard, I chase you, you run my friend, you don't go to the pub on your own, it had to be three or four

of us to go to the pub. Safety in numbers... people shout, even the people who work with you ... You get given the worst jobs ... It's upsetting, I mean, I don't say that you get used to it but what can you do? ... They say that England is a great leveller ... In 1977, I set up on my own as a builder ... by the end I must have had half a dozen people working for me. (Frank)

The patterns of socialisation the two men experienced (or aspects of their distinctive habitus) were significant for establishing gendered, racialised and classed expectations and aspirations that framed their ideas about the good life. As men in their thirties and forties Martin and Frank therefore conformed to the dominant way of living in Western societies—the notion from the Protestant Work Ethic that success in waged work and its associated affluence is the key to happiness and a good life (Weber 1990). This orientation to waged work was also a way of managing some of the insecurities they faced about their own abilities and some of the structural barriers they faced through racism and poor qualifications. This is a pattern that we have also seen with younger interviewees such as James who also became overly focused on advancement at work, bigger salaries and the consumerism this brings. What is interesting as Frank reveals is how the men assumed that their wives would share their approach to happiness—that their success at work and financial rewards would also bring happiness to their families. Of course this is not how it transpired.

I missed out on when my older children were growing up ... I was busy working ... with the benefit of hindsight it's my own fault because I was busy. Thinking that, what can I do? The best thing I can do for my family was to earn enough money to give them financial security and all this and the whole thing blew up in my face. You're so busy you know, and that in itself now I look back it's like a drug. You are there running businesses and putting an extension onto my own house, bought three bedroom houses turn them into five bedroom houses ... I thought that would keep her happy, but she wasn't. (Frank)

My eldest was 18, I was studying and I needed quiet ... he was into his heavy metal and the whole house was vibrating through his poxy music ... I went upstairs and pulled the plug out of his player and cut in into six inch

> pieces I was so angry … He was not amused and he just turned around and said, 'You care more about those kids that you're working with than you do about us' (Martin)

Towards the end of this period in their lives (their fifties) the two men began to realise there were problems with the routines they had established—their wives and children had complained about their long hours at work and little time they had for their families. They could see that their families were not contented with the consumerism and affluence of a conventional way of life. Their families also wanted time with them; they desired their attention and not just things that could be bought with the proceeds from work. Talking to me as old men, almost 20 years later, they now realise that their earlier lives had lost some balance—between work and life. Like some of the younger interviewees who we have met in this book, they too encountered some form of mid-life crisis. As research has shown, these sorts of problematic developments to biographies are very common and are difficult to monitor as individuals are fallible and struggle to assess the long-term consequences and trade-offs of small incremental changes to lifestyles (Deary 2014; Kahneman 2011; Dolan 2015). This is why these writers suggest it is so difficult to remain happy and contended in life as people often make small changes to routines that only very gradually become toxic for wellbeing. As we saw with some of the mid-life sample, both Martin and Frank had felt that they had become trapped in their way of living unable to change their lifestyles. They felt their busy routines had reduced their ability to imagine different ways of living that might be healthier and happier. Until of course some event occurred that compelled them to stop and think again about how to live.

> I got up at four in the morning having had a terrible week but I was going to a conference about this particular case and I got up in the morning, threw up, felt absolutely miserable and I absolutely come to the end of my tether and I didn't realise that. I was on the sick … got to see the doctor on the Thursday and he diagnosed it as post-traumatic stress disorder … What would I have done differently? Oh Yeah I would have paid a lot more attention, you know to my own kids and have more family activity … The worst of it was not having time, and internal conflict, how much time, you know

> to balance out things, having guilt, balancing time between my family, myself and my work. (Martin)

> I was really hurt, I was on tranquillizers, I couldn't go to work and that's why I just went to Scotland for a month, just driving around … So I was never there really, looking after the kids, I regretted that and now when I have this one, I put the family first… I have been given an opportunity, to experience something that I didn't have … I enjoy seeing this one grow, you know she's a 12 year old now and I just enjoy seeing her growing, she's my centre. (Frank)

Martin and Frank could only work at such intensities for so long before their health gave way. They both suggested they had some form of mental and physical breakdown that left them incapacitated for several months. As we saw in earlier chapters with other interviewees what seems to have occurred here is that Martin and Frank's neglect of their wellbeing did create conflicts and tensions in their psyches that eventually led to serious physical symptoms—anxiety, insomnia, mood swings and depression.

At this time in their mid-fifties, Frank and Martin were very low and just as we saw previously (in the mid-life sample) with James they struggled to feel positive about what had happened to them. However, with time this period away from waged work offered opportunity for reflection and a chance to change their routines to be happier. Martin was fortunate as his wife was in a stressful public sector job and so they were able to discuss together how best to live differently and remain together. Martin's breakdown was the catalyst for both of them to take early retirement at the age of 55 and pursue different lines of part-time work—Martin as a therapist and his wife as a yoga teacher. Whereas Frank, like so many others found that his need to rest and change career contributed to the breakdown of his marriage. Frank drew on his interest in food and bought a share in a restaurant, living again as a single man and only working part-time as a builder.[2]

[2] Lack of space precludes further discussion of later events in Frank's biography. Some of this material will be discussed in separate publications to follow.

The crises that Martin and Frank experienced in mid life, though painful at the time, gave them an opportunity to remake their lives for the better so that they could eventually, as the questionnaire data illustrates, live happier lives. The last 15 years have been a time when both men have paid more attention to the balancing of different aspects in their lives. They have avoided overworking in their waged employment and ensured they have time for family, friends, hobbies and interests that are beneficial for their wellbeing. These changes to their lifestyles have allowed them so far to remain contented despite the inevitable decline in their physical abilities and age related health problems. Remarkably, Frank has continued to flourish despite the failure of his second marriage and the financial costs associated with these breakdowns in his families. In contrast to his small rented house and old car today, a few years ago he lived in a large detached property and owned several new cars. Frank, like Martin says that he is far less materialistic now than when he was younger which may explain his happiness in much-reduced circumstances.

Aging and the Riddle of Happiness

The older people we have discussed illustrate some of the themes in this book, particularly how making sense of a good life can seem paradoxical or a riddle. Although in Western societies we often assume that prosperity is important for happiness, we found that the more affluent Martin and Mary were no happier than the poorer Frank and Jill. Though objectively the former had a better standard of living, with generous pensions, higher disposable income and home ownership this did not translate into better subjective wellbeing. The mapping exercise suggested that instead extensive social networks, interests and activities are significant for the good wellbeing of working-class interviewees. Whereas Mary's poorer wellbeing could be attributed to her older age, poorer health and more solitary lifestyle—affluence failed to compensate for these challenges. Throughout this book I have argued, however, that happiness is a complex social process that has layers of meaning and in order to discern this complexity we need to link people's accounts of their wellbeing to their changing biographies and social relationships. By doing this we were able to illuminate how and why the current activities and lifestyles of the interviewees had significance for their wellbeing.

The stories that the two female interviewees told me about their lives varied greatly and illustrate different models of how to live and pursue happiness. Mary's more materially privileged upbringing involved higher expectations of career success and an affluent married lifestyle. Though outwardly she appeared to have realised many of these conventional middle-class goals, she felt an enduring sense of disappointment about her life that marred her subjective wellbeing as she became older. Her wellbeing was conditioned by a patriarchal culture in which she was raised which offered quite narrow opportunities for middle-class women in terms of their careers, marriage and motherhood. It seems she reluctantly conformed to these expectations but always seemed aware of how she might have pursued a very different, more fulfilled life if she had lived at a different time or place. She did not have the social networks and activities that might have prevented her from dwelling on these disappointments in life. In contrast, Jill's early experiences offered her cultural resources to develop relationships that underpinned her good wellbeing throughout life. Though materially she had a very modest lifestyle, the example of her mother and role of faith and 'good works' gave her a template for how to live well. Jill's biography illustrates how good wellbeing is framed by several features such as loving family relationships, the significance of values, faith or spirituality, and working for the wellbeing of others. For Jill then happiness was not just something that you *have* but something that was also a *social practice*, that was shared, which involved everyday activities that stretched out across her community. Her kindness and compassion and the many activities these entailed in turn nourished her own wellbeing.

At face value, as many surveys would have done, we might have presumed that Martin's happiness in old age was largely attributable to his good health, happy marriage and relatively comfortable and affluent lifestyle. Undoubtedly, these are significant but his biography also suggested some complex reasons why his efforts to live well in later life had particular poignancy for him. For Frank too, the way he was concerned to organise his life in a balanced way with time for hobbies, family and friends took its meaning and significance from earlier life events. Although all of these older interviewees with good wellbeing were regularly involved in 'flow activities' it was Martin's and Frank's experience of mid-life crises and subsequent efforts to reframe their lives that gave their current life and wellbeing particular meaning. As with Mary's adoption of conventional heterosexual femininity these men too adopted a form of masculinity and associated work ethic expected of them at the time. Yet again though we see the paradox of happiness as the men developed routines that eventually undermined their wellbeing in ways that surprised and confounded them. With hindsight, both admitted to having lost a connection to important parts of their self. Marx would say that their fixation on waged work had led to their alienation from their 'species being' and the creative, imaginative selves that are essential for ones humanity and wellbeing (Marx 1984). With time, Martin and Frank were able to reflect on these crises in their lives and develop healthier and happier routes into old age. Their current happy lives have emerged out of sadness and disappointments and their efforts to manage these threats to their wellbeing. With older people what appears to be a simple, happy life can conceal a deeper, complex biography that is bittersweet or conflicted as it reflects the highs and lows of different experiences over many years. Yet many sociologists struggle to grasp this complexity of how happiness emerges across a lifetime out of the efforts people make to live well with their loved ones. There is a tendency instead to regard happiness more pessimistically as a simpler, emotional or psychological state shaped by structures over which individuals have little control.

Bibliography

Ahmed, S. (2010). *The promise of happiness*. London: Duke University Press.

Aristotle. (2009). *Nicomachean ethics*. Oxford: Oxford University Press.

Bernard, M., Phillips, J., Machin, L., & Harding Davies, V. (Eds.) (2000). *Women aging: Changing identities, challenging myths*. London: Routledge.

Blanchflower, D., & Oswald, A. (2008). Is wellbeing U shaped over the life cycle? *Social Science and Medicine, 66*, 1733–1749.

Blane, D., Netuvelli, G., & Bartley, M. (2007). Does quality of life at older ages vary with socio-economic position? *Sociology, 41*(4), 717–726.

Bourdieu, P., & Passeron, J. C. (1990). *Reproduction in education, society and culture*. London: Sage Press.
Bowling, A., & Dieppe, P. (2005). What is successful aging and who should define it? *British Medical Journal, 331*, 1548.
Burke, E. (2015). *A philosophical enquiry into the origin of our ideas of the sublime and beautiful* (2nd ed.). Oxford: Oxford University Press.
Cieslik, M. (1997). *Youth, disadvantage and the underclass in South Wales*. Unpublished PhD Thesis, University of Kent at Canterbury.
Cieslik, M., & Simpson, D. (2015). Basic skills, literacy practices and the hidden injuries of class. *Sociological Research Online, 20*(1). Retrieved from http://www:socresonline.org.uk/20/1/7.html
Cooper, L., & Thomas, H. (2002). Growing old gracefully—Social dance in the third age. *Aging and Society, 22*(6), 689–708.
Csikszentmihalyi, M. (2002). *Flow: The classic work on how to achieve happiness*. London: Harper and Row.
Csikszentmihalyi, M., & Leferve, J. (1989). Optimal experience at work and leisure. *Journal of Personality and Social Psychology, 56*(5), 815–822.
Deary, V. (2014). *How to live*. London: Allen Lane.
Dolan, P. (2015). *Happiness by design: Finding pleasure and purpose in everyday life*. London: Penguin.
Goffman, E. (1956). *The Presentation of Self*. Edinburgh: Edinburgh University Press.
Green, L. (2010). *Understanding the life course: Sociological and psychological perspectives*. Cambridge: Polity Press.
Hochschild, A. (2003). *The managed heart: The commercialisation of human feeling* (2nd ed.). Berkeley: University of California Press.
Hoggard, L. (2005). *How to be happy*. London: BBC Books.
Kahneman, D. (2011). *Thinking fast and slow*. London: Penguin Books.
Kant, I. (1951). *Critique of judgement*. London: Macmillan.
Kant, I. (2003). *Observations on the feeling of beautiful and sublime*. London: University of California Press.
Lawton, M. P., Winter, L., Kleban, M. H., & Ruckdeschel, K. (1999). Affect and quality of life: Objective and subjective. *Journal of Aging and Health, 11*, 16–198.
Layard, R. (2005). *Happiness: Lessons from a new science*. London: Penguin.
Marx, K. (1984). *Marx: Early writings*. Harmondsworth: Penguin.
Oakley, A. (1974). *The sociology of housework*. Oxford: Martin Robertson.
ONS. (2015a). *Measuring national wellbeing: Life in the UK 2015*. London: ONS.

ONS. (2015b). *Measuring national wellbeing. Insights into loneliness, older people and wellbeing, 2015*. London: Office for National Statistics.

Pahl, R. (2007). Friendship, trust and mutuality. In J. Haworth & G. Hart (Eds.), *Wellbeing: Individual, community and social perspectives* (pp. 256–270). London: Palgrave.

Scambler, S., Victor, C. R., Bond, J., & Bowling, A. (2002). *Promoting quality of life: Preventing loneliness amongst older people*. Paper presented to XV World Congress of Sociology, 7–13 July, Brisbane, Australia.

Schiller, F. (2004). *Letters on the aesthetical education of man*. New York: Dover Publications.

Thin, N. (2012). *Social happiness: Theory into policy and practice*. Bristol: Policy Press.

Van den Hoonaard, D. (2007). Editorial: Aging and masculinity: A topic whose time has come. *Journal of Aging Studies, 21*, 277–280.

Vincent, J. (2003). *Old age*. London: Routledge.

Ward, L., Barnes, M., & Gahagan, B. (2012). *Wellbeing in old age: Findings from participatory research*. Brighton: University of Brighton.

Weber, M. (1990). *The protestant ethic and the spirit of capitalism*. London: Allen Unwin.

11

Conclusions: Making Sense of the Happiness Riddle

This project began in the early years of this century as it seemed that apart from lone figures such as Ruut Veenhoven, mainstream sociology neglects some of the key things that make us human—our desire to be happy and live well. For a discipline that prides itself on its ability to study just about anything it was paradoxical that most contemporary sociologists had ignored such fundamental issues. Although during the course of this project there has been some new work in the sociology of happiness (see Hyman 2014; Jugureanu et al. 2014), this area is still under-researched and requires much more engagement from sociologists in Britain and elsewhere.

In this book I suggest that sociologists tend to be sceptical about the value of happiness research as happiness is often narrowly conceived as a subjective, positive experience. Such notions of happiness are then portrayed as tools of corporate interests and elites, used as marketing devices to promote consumerist lifestyles (Davies 2015). Happiness is seen to function as a handmaiden of capitalism and Neo-Liberal governments, nudging us to worry more about how we look and feel than the economic, cultural and political processes that structure our life chances (Furedi 2004; Ahmed 2010). For many sociologists, therefore,

M. Cieslik, *The Happiness Riddle and the Quest for a Good Life*,
DOI 10.1057/978-1-137-31882-4_11

researching happiness is a distraction from the more important analysis of the deeper causes of social division, domination and inequality that characterise our societies. Modern sociologists perhaps feel justified in pursuing this approach as they are following in the footsteps of the founding figures in the discipline and those such as the Frankfurt School that followed them. To research happiness sociologically does entail a questioning or an, 'unlearning' of some of the founding tenets of the discipline. It also incurs some resistance and hostility, as there is the perception that by studying happiness we are supporting those reactionary ideas that underpin consumer capitalism.

My interest in happiness was piqued as academic disciplines such as economics and sociology represent it in narrow ways that seemed at odds with my own experiences. As we have seen with the participants in this research happiness can be an imaginative process—something that we do in our heads—remembering, reflecting and anticipating (see Hyman 2014). This creativity we rarely see in contemporary analyses of wellbeing such as economics with its focus on life satisfaction or with sociology and its interest in happiness scripts (Ahmed 2010) and narcissism (Furedi 2004).

Happiness involves a landscape of different emotions and displays that are part of the fabric of life, yet seldom is this richness explored by sociologists who are more concerned with the pathologies of life (Hochschild 2003). But working at 'feeling good', psyching oneself up, 'being positive' are all important experiences and skills we develop and should be acknowledged and researched.

Living well and seeking happiness are also inherently social and political processes. As we have seen, happiness has meaning as we pursue it with others, for our own personal happiness, though important, is greatly amplified if our loved ones can also live well. At various times we find our efforts to be happy conflicts with others—partners change, we change, jobs change and we find ourselves struggling to live differently so we can still be happy. The interviewees were all challenged in these ways to monitor their wellbeing as they got older and some managed better than others as they had greater insight, more resources or better networks. In many different settings we see people pursuing their interests oblivious to the wellbeing of others—their success and happiness comes with the cost of misery to many other people. Some have the power and resources to live well but what sort of happiness is it that also leads to sadness, suffering and loss for so many others? Happiness, therefore, raises questions about fairness, justice, the ethics of how people live and the sorts of values that inform these features of life (Sayer 2011).

As the scepticism about happiness studies is ingrained in sociology, I have been forced to look to philosophy and other traditions in psychology, economics and psychoanalysis to help in the development of a sociologically informed study of happiness. My approach is necessarily cross-disciplinary as it synthesises insights from different traditions of wellbeing research. The aim has been to develop a far more ambitious engagement with the nature of happiness than we usually see with sociology. The interest that sociologists have with the pathologies of modern life means they often close down questions about how we experience happiness and how wellbeing functions in our daily lives. As a result, modern sociology by documenting the suffering of people often seems to create unbalanced accounts of life when compared to how people actually live their lives.

Darrin McMahon's book on the History of Happiness (2006) illustrates the very complex way that classical philosophers engaged with the nature of wellbeing and demonstrates how crudely many sociologists regard the subject. From these philosophers I took the notion of happiness as being a social and collaborative practice, rather than just as a subjective, emotional experience—'pleasure seeking' as sociologists often view it. This book, therefore, has been about exploring how people work at their happiness, with other people, balancing the good and not so good events over their life course. From economists such as Layard (2005), Blanchflower and Oswald (2008), I took the many different insights about the sources of happiness such as income, aging, family relationships and so on and explored the significance of these for my interviewees. Psychologists have researched the many ways that people do have positive experiences in life (Csikszentmihalyi 2002), and these have been useful as a corrective for the usually pessimistic way that sociologists often regard work, employment, adolescence, families and so on. Psychologists and psychoanalysts also show us how we misremember the past and have cognitive biases so that we struggle to make informed choices about our happiness. The notion of fallibility has been an important one in interviews, where I have explored with respondents about how they make choices about living well.

Although the insights from other disciplines have helped to correct some of the deficiencies in sociological research into happiness, at the same time there are also weaknesses in how these other approaches study wellbeing. Psychology, economics, psychoanalytical theory and philosophy tend to neglect the way that power relationships and social divisions structure people's lives and their ability to be happy. They often use experiments and abstract models of people and different scenarios to explore wellbeing that are far removed from the messy, shifting dynamics of how people actually live. Hence, despite my criticisms of mainstream sociology, its traditions do offer valuable tools that I have used in this project. In particular, in the way that sociologists connect economic, political, cultural or policy processes to the opportunities and resources of individuals which in turn frame their life experiences (Mills 1959). I have tried to account for the differences in my interviewees' wellbeing by relating their happiness to this interplay of structures and agency over the individual life course. Some people in the research had more opportunities and greater resources than others (which have been linked to patterns of gender, class, race and sexuality) that then had a bearing on their experiences of happiness. Yet these predictable accounts of the structuring of wellbeing are also joined by others that appear more puzzling—for sociologists at least. There were many instances where people's wellbeing could not always be read off from the resources or opportunities they enjoyed. The secrets of their happiness were found only after lengthy analysis of their life course transitions and the influence of many different relationships that they had over many years. And here then we see some of the paradoxical features of happiness, which reminds us of the need for a cross-disciplinary approach to the riddle of happiness.

The Structuring of Happiness?

Why then were some people happier than others? When I analyzed the opportunities and resources of interviewees, one could see how their wellbeing might be linked to those important factors identified by economists and psychologists in their discussions of happiness and a good life, such as good health, income, relationship, family, values, work and so on (Layard 2005: 63). There were relationships between objective circumstances and the feelings and emotions of interviewees about their wellbeing. For my interviewees there were benefits to having good health, being loved, a long-term partnership, rewarding employment and a reasonable income. I used some concepts from psychologists such as Csikszentmihalyi (2002) or Marx's ideas around non-alienation (1984) to unpack how work or leisure activities can be nourishing as they offer opportunities for learning, the development of skills, creativity, friendships and links with nature.

However, existing survey research focuses on the wellbeing of average people in abstract contexts, so we get little sense of how these different factors are experienced and how they may interact to influence wellbeing in specific social and historical contexts. By employing a biographical approach I was able to tease out some of these processes and examine the patterning of wellbeing in relation to age, class and gender and how different people also tried to manage and adapt to the structuring of their wellbeing. The young woman Teresa, for example, was unable to live independently as her job was low paid whilst housing costs were relatively high. The long-term dependency on her parents was a threat to her wellbeing—one that is common to many young people in the UK. Yet at the same time Teresa had an extensive friendship network, good relations with parents and a boyfriend which all supported her wellbeing. This conjunction of threats and resources meant that young people's wellbeing tends to 'ebb and flow' as they made their transitions to adult independence. In contrast, Louise had much poorer wellbeing than Teresa because of the corrosive effects of family breakdown and parental drug use. Louise then was less able to manage some of the material insecurities of working-class adolescence because of some of the trauma she experienced when young.

I examined some of the influences of ageing on wellbeing such as whether people experience a mid-life crisis. Although the middle-class participant James appeared to have experienced a crisis in mid-life further analysis revealed a longer process of poor wellbeing, stemming from events in his thirties. He had devoted much of his time to career advancement and this lack of balance in his life (and neglect of his family and partner) had contributed to later problems with his wellbeing. The others in the mid-life sample all tended to have had 'mini crises' over a longer period of time rather than one centred on mid-life. Interestingly, a higher income did not guarantee happiness for the mid-life sample as those with better wellbeing tended to be the ones who worked less and had a more balanced life. But these different biographies do raise questions about how we judge a successful life. James, for example, had achieved much—success at work, large house, expensive cars and several children—yet also a failed marriage and unhappiness for many years? There appeared to be a mismatch between his objective circumstances and his subjective wellbeing.

I also explored transitions into long-term relationships and parenting that the Thirty Somethings were making—a key phase in the life course. Here we saw examples of how happiness can function as 'a problem' (Ahmed 2010) as interviewees were struggling to achieve lifestyles goals. In this sample we also witnessed the gendering of experiences that shaped wellbeing in distinctive ways. Some of the women's opportunities to flourish were hindered by the controlling behaviours of male partners. This was one example of how happiness is interwoven with power relationships and involves trade-offs and struggle between different family members. This reminds us of the way happiness is a social process rather than a characteristic of individuals or a simple emotional state as 'good feeling'.

With the sample of older interviewees we had an insight into some of the ways that good wellbeing is achieved in old age. For Frank and Martin their earlier difficulties in life ensured they worked hard to live well in retirement and so their interests in food and art had particular poignancy for them. With the older women too we were able to relate their wellbeing to earlier life events. Jill, for example, enjoyed better wellbeing because of lifelong interests and networks, whereas Mary had poorer wellbeing linked to her social isolation.

A key feature of interviewees' accounts of happiness was when people were surprised by their emotional responses to what seemed innocuous events. I suggested that some psychoanalytical theory would be useful to interpret these puzzling aspects of biography and wellbeing. Alice, in the young people's sample, experienced such an event when on a gap year prior to university. In the interview we worked through these events, suggesting that the negative responses she experienced were linked to earlier life experiences. In other interviewees, such as Greg as we saw with his efforts to manage disability, early life events (such as nurturing relations with parents) can promote later resilience. The insight from these examples was how sometimes people are unaware of how wellbeing is structured—they are fallible, unaware of how experiences can trigger surprising responses that influence their wellbeing. This is one way in which happiness is often experienced as a riddle or enigma.

Some interviewees spoke of good or bad fortune shaping their wellbeing in ways that the Greek philosophers would recognise and show how chance or luck has a role to play in happiness. Greg had the misfortune to suffer a brain injury that transformed his life, greatly limiting his opportunities to flourish. When interviewed, he was still trying to come to terms with his disability and these changes that posed serious threats to his wellbeing. Paula by contrast, through a chance encounter, was able to escape an abusive relationship and make a new life for herself with a new partner and new home. This illustrates the significance of love for good wellbeing (Thin 2012). It also suggests that even when wellbeing appears to have been structured by class, gender or age a 'fateful moment' can intervene to reshape a life and happiness in surprising ways.

What Can We Do to Be Happier?

Some sociologists (Marcuse 2002; Ahmed 2010) are pessimistic about the power that individuals have to improve their wellbeing. Classical sociologists suggest that wellbeing is structured by social networks (Durkheim 2014), relations of production (Marx 1984) or systems of belief and meaning (Weber 1990) rather than through the activities of individuals. Whereas some psychologists and self-help specialists (Williams and Penman 2011; Ben-Shahar 2008) argue that people have the potential for positive change and to work at being happier, the life stories in this book suggest a middle ground between these two positions. People can make choices and changes that can improve their wellbeing but other people, past events and power relations working through class, gender and 'race' all influence happiness.

If we compare countries that score highly in wellbeing surveys, such as Denmark, Sweden and the Netherlands with the UK which does less well we can see some public policy initiatives that the UK could develop to enhance wellbeing (see Bacon et al. 2010; Christie and Nash 1998). These involve the expansion of affordable housing that has become such a problem for young people in the UK today. These 'happier societies' also have family friendly policies that offer more childcare, flexible working and better support for parents allowing families to enjoy more time together away from the pressures of work (McDonald and Kragh 2011). These countries have also invested in integrated transport networks that encourage cycling, walking and public transport use that all have long-term health and wellbeing benefits. These policies also improve the quality of life in communities by reducing the corrosive effects of car use and congestion on health and wellbeing. Happier societies also tend to have more generous welfare provision than seen in the UK for poverty is a key barrier to good wellbeing. Similarly, good provision for children (in terms of education, health and social services) is essential to help nurture them to become productive and happy adults (Layard and Dunn 2009). These happier societies tend to have progressive tax regimes that reduce inequality and generate the revenue to fund these policies that can underpin good wellbeing. Yet in the UK, since the 1980s consecutive governments have pursued a 'quasi-markets' approach to social policy that has promoted privatisation and a laissez faire role for government that has undermined rather than promoted good wellbeing.

Despite the structural constraints on wellbeing in the UK, we did see various examples of people making concerted efforts to live happier, more rewarding lives. Many of these strategies involved trying to develop better work-life balance, acknowledging the need for work and income but also time for hobbies, friends and family. Nigel and Kate in the mid-life sample drew on their interests and values around socialism and environmentalism to develop 'alternative' ways of living that challenged the usual consumerist lifestyles we see today. People had developed ways to experience 'flow' (Csikszentmihalyi 2002) or Marx's notion of flourishing (Marx 1984) in their lives ensuring a variety of nurturing events—relationships with others, time in nature, opportunities to learn and develop skills and chance to create things of use. Gardening on an allotment was a classic example of an activity that fulfilled many of these features of flourishing. Others had developed interests in music, art, food and dance that aided their happiness.

Other interviewees such as Alice and Jill were both religious and their beliefs in 'good works' and compassion helped them to flourish. As we have seen earlier many of the people I interviewed with better wellbeing relied on a structure of beliefs, values or spirituality to make sense of their life and their place in the world. Being able to live with meaning was central to the happiness of my interviewees. James for example, after a difficult divorce, had used mindfulness techniques (Williams and Penman 2011) to improve his wellbeing and this offered a set of beliefs and values that promoted better living. It offered him tools to challenge some of the instrumentalism that had governed his life and how to explore a more balanced lifestyle.

The culture of most Western societies is one that celebrates speed, materialism, consumption, hard work and career success as markers of a successful life. Thus, to work less, to live more slowly and in a more balanced way with time for friends, family and hobbies takes courage. Cultures at work, the media and politicians exhort us to work harder, faster, further and so as Aristotle reminds us, every day we have to work a little to live a good life.

Bibliography

Ahmed, S. (2010). *The promise of happiness*. London: Duke University Press.

Bacon, N., Brophy, M., Mguni, N., Mulgan, G., & Shandro, A. (2010). *The state of happiness: Can public policy shape people's wellbeing and resilience?* London: Young Foundation.

Ben-Shahar, T. (2008). *Happier*. London: McGraw Hill.

Bentham, J. (1776). *'Preface to the first edition', A fragment on government*. In Mack, M. P. (Ed.). (1969). *A Bentham reader*. New York: Pegasus Press, p. 45. In McMahon, D. (2006). *Happiness: A history*. New York: Grove Press, p. 212.

Blanchflower, D., & Oswald, A. (2008). Is wellbeing U shaped over the life cycle? *Social Science and Medicine, 66*, 1733–1749.

Christie, I., & Nash, L. (1998). *The good life*. London: Demos. Retrieved from http://www.demos.co.uk/files/thegoodlife.pdf?1240939425

Csikszentmihalyi, M. (2002). *Flow: The classic work on how to achieve happiness*. London: Harper and Row.

Davies, W. (2015). *The happiness industry: How the government and big business sold us wellbeing*. London: Verso.

Durkheim, E. (2014). *The division of labour in society*. New York: Free Press.

Furedi, F. (2004). *Therapy culture: Cultivating vulnerability in an uncertain age*. London: Routledge.

Hochschild, A. (2003). *The managed heart: The commercialisation of human feeling* (2nd ed.). Berkeley: University of California Press.

Hyman, L. (2014). *Happiness: Understandings, narratives and discourses*. London: Palgrave.

Jugureanu, A., Hughes, J., & Hughes, K. (2014, May 2). Towards a developmental understanding of happiness. *Sociological Research Online, 19*.Retrieved from http://www.socresonline.org.uk/19/2/2.html10.5153/sro.3240

Layard, R. (2005). *Happiness: Lessons from a new science*. London: Penguin.

Layard, R., & Dunn, J. (2009). *A good childhood: Searching for values in a competitive age*. London: Penguin.

Marcuse, H. (2002). *One dimensional man: Studies in the ideology of advanced industrial society*. London: Routldege.

Marx, K. (1984). *Marx: Early writings*. Harmondsworth: Penguin.

McDonald, L., & Kragh, N. (2011, January 1). Family friendly policies in Denmark. *Guardian Newspaper*.

Mills, C. W. (1959). *The sociological imagination*. Oxford: Oxford University Press.

Sayer, A. (2011). *Why things matter to people: Social science, values and ethical life*. Cambridge: Cambridge University Press.

Weber, M. (1990). *The protestant ethic and the spirit of capitalism*. London: Allen Unwin.

Williams, M., & Penman, D. (2011). *Mindfulness: A practical guide to finding peace in a frantic world*. London: Piatkus.

Bibliography

Beder, S. (2001). *Selling the work ethic: From Puritan pulpit to corporate PR.* London: Zed Books.

Bradshaw, J. (2015). Subjective wellbeing and social policy: Can nations make their children happier? *Child Indicators Research, 8*(1), 227–241.

Brown, P. (1987). *Schooling ordinary kids: Inequality, unemployment and the new vocationalism*. London: Tavistock.

Chancer, L., & Andrews, J. (Eds.) (2014). *The unhappy divorce of sociology and psychoanalysis: Diverse perspectives on the psychosocial.* London: Palgrave.

Charles, N., & Harris, C. (2007). Continuity and change in work-life balance choices. *British Journal of Sociology, 58*(2), 277–295.

Choe, J., Chick, G., & O'Regan, M. (2015). Meditation as a kind of leisure: The similarities and differences in the United States. *Leisure Studies, 34*(4), 420–437. doi:10.1080/02614367.2014.923497.

Clark, T. (2014). *Hard times: The divisive toll of the economic slump*. Yale: Yale University Press.

DfE. (2014). *Outcomes for children looked after by local authorities in England as of 31st March 2014*. London: DfE.

Edgell, S. (2012). *The sociology of work*. London: Sage.

Feltham, C., & Dryden, W. (Eds.) (1992). *Psychotherapy and its discontents.* Buckingham: Open University Press.

M. Cieslik, *The Happiness Riddle and the Quest for a Good Life*,
DOI 10.1057/978-1-137-31882-4

Frayne, D. (2015). *The refusal of work: The theory and practice of resistance to work*. London: Zed Books.

Gellner, E. (1992). Psychoanalysis, testability and social role. In C. Feltham & W. Dryden (Eds.), *Psychotherapy and its discontents* (pp. 41–51). Buckingham: Open University Press.

Golding, R. (1982). Freud, psychoanalysis and sociology: Some observations on the sociological analysis of the individual. *British Journal of Sociology, 33*(4), 545–562.

Guardian. (2015). *The housing trap: How can Berlin avoid following in London's pricey footsteps?* Wednesday 23rd September. Retrieved from http://www.theguardian.com/cities/2015/sep/23/housing-trap-how-berlin-avoid-following-london-pricey-footsteps

Joseph Sirgy, M., & Wu, J. (2009). The pleasant life, the engaged life and the meaningful life: What about the balanced life? *Journal of Happiness Studies, 10*, 183–196.

MacDonald, R., Kreutz, G., & Mitchell, L. (Eds.) (2012). *Music, health and wellbeing*. Oxford: Oxford University Pres.

MacDonald, R., & Shildrick, T. (2013). Youth and well-being: Experiencing bereavement and ill-health in marginalised young people's transitions. *Sociology of Health and Illness, 35*(1), 147–161.

Metz, C. (1994). *Psychoanalysis and cinema: The imaginary signifier*. London: Palgrave.

Montserrat, C., Dinisman, T., Bălţătescu, S., Grigoraş, B. A., & Casas, F. (2014). The effect of critical changes and gender on adolescent's subjective wellbeing: Comparisons across 8 countries. *Child Indicators Research, 8*(1), 111–131.

Myrdal, A., & Klein, V. (1956). *Women's two roles: Home and work*. London: Routledge and Kegan Paul.

Rees, G., & Dinisman, T. (2014). Comparing children's experiences and evaluations of their lives in 11 different countries. *Child Indicators Research, 8*(1), 5–31.

Roberts, K. (2011). Leisure: The importance of being inconsequential. *Leisure Studies, 30*(1), 5–20. doi:10.1080/02614367.2010.506650.

Siegel, L. (2005, May 8). Freud and his discontents. *The New York Times*.

The Children's Society. (2015). *The good childhood report*. London: Children's Society.

Veblen, T. (2009). *The theory of the leisure class*. Oxford: Oxford University Press.

Index

Note: Page numbers followed by n denote foot notes.

M. Cieslik, *The Happiness Riddle and the Quest for a Good Life*,
DOI 10.1057/978-1-137-31882-4

E

F

G

M

N

O

P